DIRECTORY OF TOXICOLOGICAL AND RELATED TESTING LABORATORIES

DIRECTORY OF TOXICOLOGICAL AND RELATED TESTING LABORATORIES

Regulatory Assistance Corporation
Havenwell Junction, New York

◉ HEMISPHERE PUBLISHING CORPORATION
A member of the Taylor & Francis Group
New York   Washington   Philadelphia   London

DIRECTORY OF TOXICOLOGICAL AND RELATED TESTING LABORATORIES

Copyright © 1991 by Hemisphere Publishing Corporation. All rights reserved. Printed in the United States of America. Except as permitted under the United States Copyright Act of 1976, no part of this publication may be reproduced or distributed in any form or by any means, or stored in a database or retrieval system, without the prior written permission of the publisher.

1 2 3 4 5 6 7 8 9 0   BR BR   9 8 7 6 5 4 3 2 1

This book was set in Times Roman by Hemisphere Publishing Corporation. The editors were Michele Nix and Heather Jefferson; the production supervisor was Peggy M. Rote; and the typesetter was Phoebe Carter. Cover design by Debra Eubanks Riffe. Printing and binding by Braun-Brumfield, Inc.

A CIP catalog record for this book is available from the British Library.

Library of Congress Cataloging-in-Publication Data

Directory of toxicological and related testing laboratories /
   Regulatory Assistance Corporation.
       p.   cm.
       Includes bibliographical references and index.

     1. Toxicology laboratories—United States—Directories.
   I. Regulatory Assistance Corporation.
RA1193.7.D56   1991   90-25635
604.7—dc20   CIP
ISBN 0-89116-904-0

Contents

Preface	vii
Acknowledgments and Listings	ix
Using This Directory	xi
Introduction: Selecting a Toxicology Laboratory	xiii
Part I: Listing of Toxicology Laboratories	1
Part II: Appendices	83
Appendix A: Summary List of Laboratory Contacts	85
Appendix B: List of Studies Conducted and/or Services Provided	91
Appendix C: Types of Chemicals Tested	95

Preface

Growing public and governmental concerns with the health, safety, and environmental effects of chemicals has resulted in a tremendous increase in the need to test new and existing chemicals in order to determine the hazards and risks associated with their manufacture, distribution, and use.

As a result, companies are increasingly concerned with locating laboratories that have the experience and expertise required to conduct reliable and acceptable studies for use in either company sponsored programs or in compliance with a number of government agency regulations.

This directory of U.S. toxicology, ecotoxicology, environmental, analytical, and support service laboratories is specifically designed to meet this need. While there have been directories of toxicology laboratories published in the past, none have been issued recently. Those that were provided information on toxicology testing laboratories only. They did not attempt to provide any extensive coverage of ecotoxicology or environmental testing laboratories. Even more importantly, they did not provide listings of laboratories that could provide required analytical support or specialized services, such as the synthesis of radio labeled material required for certain studies.

This volume was designed to fulfill these needs and to provide, in one compact resource volume, a directory with useful and concise information on the following:

- Toxicology Laboratories (mammalian and related studies)
- Ecotoxicology (aquatic and related studies)

- Environmental Effects/Fate
- Analytical Services (methods development and analysis)
- Specialized Services, i.e., Radio Label Compound Synthesis

As indicated above, this directory only includes laboratories actually conducting tests in the United States.

Acknowledgments and Listings

Regulatory Assistance Corporation (RAC) wishes to acknowledge and thank all of those companies that provided the information requested so that the firms could be listed in this directory.

While every effort was made to be as all inclusive as possible, it must be noted that several laboratories specifically declined to be listed whereas others for whatever reason did not respond to repeated inquiries. Therefore, it is recognized that there are some omissions but these are completely beyond our control or as a result of an unintentional omission.

Future editions of this directory are planned and, therefore, any laboratory not listed but which does wish to be included in the future should contact: Regulatory Assistance Corporation, 17 Clearview Circle, Hopewell Junction, New York 12533.

Using This Directory

This directory is divided into two basic parts:

Part I consists of a summary of the following information for each laboratory included in alphabetical order by the name of the laboratory:

Name of company
Address
Corporate officers
Services offered
Tests performed
Products tested
Special expertise
Size of laboratory
Equipment available
Key personnel
Staff
GLP compliance
FDA approval
Accreditation
Communications capabilities
Additional data
Key contact

Part II consists of three appendices that will enable the user to locate candidate laboratories either by name (Appendix A in alphabetical listing), the type of study required (Appendix B, which lists studies conducted and/or services provided), or by types of chemicals tested (Appendix C).

Appendix A is a summary table of all laboratories listed alphabetically including address, telephone number, telefax number, telex number, and key contact to be used as a convenient reference for each laboratory.

Appendix B is an alphabetical list of studies conducted and/or services provided by individual laboratories:

Acute
Analytical Services
Aquatic
Chronic
Cytotoxicity
Dermatology Efficacy
Developmental Toxicology & Teratology
Ecotoxicology
Environmental Fate
Genotoxicity
Immunotox
Industrial Hygiene/Analytical
Inhalation Toxicology
Metabolism
Microbiology
Neurotoxicology (Includes Behavioral and Neurobehavioral)
Pharmacokinetics/Toxicokinetics
Radiolabeled Synthesis
Reproduction
Special Studies in Toxicology
Subchronic

Introduction: Selecting a Toxicology Laboratory

The selection of a toxicology laboratory is critical for several reasons. Toxicology and ecotoxicology tests and related services can be expensive, therefore there is a definite economic consideration. However, even more important than any monetary consideration is the critical aspect of the eventual acceptability of the study and the availability of the study results on schedule. The latter point, namely scheduling and timing, is important whether or not the study or studies are being conducted for the company's own purposes or in compliance with specific government regulations.

Those who have never been involved in the laboratory selection process might find it helpful to consider the following factors and recommendations. Clearly the amount of time and the resources dedicated to laboratory selection will depend upon the nature and the extent of the studies that are going to be conducted. Selecting a qualified laboratory to conduct comparatively simple acute studies is far less critical and involved than the selection of a laboratory to conduct a long-term chronic bioassay or a series of several assays. Therefore, this should be taken into consideration when assessing the need to pursue all of the following recommendations and considerations.

Recommendations for Selecting a Laboratory

The following list is provided as a guide to assessing a laboratory's qualifications and as a basis for selecting the laboratory:

1. Obtain all available literature from the laboratory. This should include a description of the facility, information on the nature of the studies conducted, as well as a statement of the laboratory's experience and qualifications.

2. Obtain copies of the CVs of the laboratory's principal personnel.

3. If possible, discuss the laboratory's reputation with others who have already used the laboratory. Specifically, determine the users' opinions of the laboratory in the context of: The types of studies to be placed at the laboratory; professional experience and expertise in the disciplines involved; availability of laboratory personnel for discussions and study review; and timeliness of study completion and issuance of reports.

4. Obtain copies of the protocols of any studies planned. In some cases the protocols are standardized and readily available. In other cases it may be necessary to develop a specialized protocol. Usually a basic "model protocol" can be provided, even though it might require modification in the context of the chemical properties or some particularized study requirements and objectives.

5. Obtain a study schedule—that is, when can the study be initiated and what is the expected date of completion. When considering the date of completion, remember that many studies are divided into distinct phases. Study completion will normally mean the date at which all of the experimental aspects of the study have been completed, but it does not mean the date of either a draft or final report. This is an important distinction since it often takes several months between the time of completing the experimental portion of the study and the availability of even a draft report. Therefore, obtain a schedule that precisely indicates the completion dates of all key phases including the scheduled availability of interim, draft, and final reports.

6. Obtain price quotations. Again, there can be substantial variations in cost estimates from laboratory to laboratory so be certain that any cost quotations are in fact comparable. That is, be sure that the quotations are being made on the basis of the same study design and protocol.

7. Analytical support costs are usually not included in study cost quotations. Many studies, i.e., subchronic, chronic, etc., require considerable analytical support. Analytical support charges are normally quoted separately. Always determine if analytical support is required and whether or not any cost estimates for such analytical support are included in the study cost quotation. If they are not included in the study cost quotation, obtain an estimate, because analytical costs can represent a substantial percentage of the total study cost.

8. Range finding studies—that is, preliminary studies to determine appropriate dose levels—are required for many assays, i.e., subchronic and chronic, reproduction, and teratology. The range finding studies are usually not included in cost estimations nor in the protocols if a quotation on the basic study is requested without specifying whether or not a range finding study is required. Therefore, determine whether or not a range finding study is necessary. Where it is necessary, obtain a cost and protocol for the range finding study as well as the study itself.

9. Wherever possible, and assuming that it is cost justified, visit the site of any candidate laboratories before placing studies with them. The following "check list" contains several criteria that should be utilized in evaluating a laboratory or in comparing two or more laboratories. It can be taken and utilized during site visits. Of course, in those instances where actual site visits are impossible or not cost justified, the check list can still be used based upon an assessment made through the company's literature, personnel CVs, and discussions.

10. Secure a written contract with the laboratory. Most laboratories have their own standard contract forms that specify the applicable terms and conditions. Review these carefully and if necessary modify them or develop separate contracts.

Check List for Laboratory Selection

The following check list should be used to evaluate an individual laboratory or to compare two or more laboratories. In the event that a site visit is possible, these questions should be asked during the visit. If a site visit is either impractical or not economically justified, this information should be obtained from a review of the laboratory's prepared literature, as well as through discussion with the laboratory representatives.

- Is the lab accredited?
- What specific accreditation does it have?
- Does the lab perform tests in compliance with Good Laboratory Practice (GLP) requirements?
- Has the lab received a GLP Compliance inspection; if so: By what agency(ies)? When was the inspection conducted? What was the result of the inspection?
- Has the laboratory undergone any study audits; if so: By what agency(ies)? What type of studies were audited? When were the audits conducted? What were the results of the audits?
- Does the laboratory have standard operating procedures, and are these written and available for inspection?
- Are CVs for all personnel available and have personnel the appropriate education, training, or certification required appropriate to their particular responsibilities?
- Does the laboratory actually have the equipment necessary to conduct the tests?
- How physically accessible is the laboratory—how convenient or inconvenient is it to visit the laboratory?
- What communication facilities are available, i.e., telephone, telefax, or on-line computer access?
- Are operations computerized?
- Will all services actually be conducted by the laboratory itself or will there be any subcontracting involved?
- What personnel safety programs and training are available?
- Has the laboratory ever been cited for any violations, i.e., TSCA, OSHA, etc.?
- What simple identification and tracking system is employed?
- How are samples stored?
- What is the laboratory's archival capability?
- How is any hazardous waste stored, treated, or disposed?
- Is the laboratory able to use radio labeled materials?
- Does the laboratory have necessary permits either federal, state, or local?
- Has the laboratory actually conducted the type of study(ies) involved?
- What animal care and handling facilities are available?
- Have any of the laboratory's studies ever been rejected by federal or state agencies? If so, why?
- What is the laboratory's present capacity versus utilization rate?
- Will the laboratory permit sponsor or sponsor representative site visits?
- Will the laboratory permit the use of a third party study auditor?
- How are reports prepared, i.e., are they on computer and therefore easily modified and generated?
- What are the laboratory's standard contract terms and conditions?
- Is the laboratory knowledgable of specific regulations that might be involved in the conduct of a given study?

- Does the laboratory have actual experience with complying with any federal or state regulations that might be involved?
- Does the laboratory have any on-going working relationships with other laboratories that might be needed to support the program?
- Does the laboratory have on-staff or readily accessible veterinary care?
- Does the laboratory do its own histopathology and does it have its own histopathologists?

Once the above information is obtained it can be used to assess a given laboratory or to compare the capabilities of two or more laboratories that are being considered for the study program.

Part I

Listing of Toxicology Laboratories

1. **Name of company:**
   American Health Foundation
2. **Address:**
   1 Dana Road
   Valhalla, New York 10595
3. **Corporate officers:**
   E. L. Wynder, M.D., President
   G. M. Williams, M.D., Director of Medical Sciences
   S. S. Hecht, Ph.D., Director of Research
   R. B. Klarberg, Esq., Executive Vice President
   H. Taylor, Chairman, Board of Trustees
4. **Services offered:**
   Analytical chemistry, organic chemicals MS, NMR, GC, HPLC, UV, IR, Thermoelectron, isotopically labeled chemicals. Toxicology, chronic in rodents. Genotoxicity, promoters, in vitro systems approach, risk assessment, serum cholesterol.
5. **Tests performed:**
   Chronic toxicity, in vitro and in vivo genetic toxicology promoters, rodent species, Ames tests, Williams tests, cell communication, serum cholesterol, nutrition, health.
6. **Products tested:**
   Any type of organic chemicals.
7. **Special expertise:**
   Genetic toxicity in Salmonella systems (Ames) or in vitro/in vivo DNA repair in liver cells (Williams); cell to cell communication; histopathology, diagnostic pathology; organic chemical separation techniques, nitrosamines, PAH, analyses.
8. **Size of lab:**
   60,000 square feet.
9. **Equipment available:**
   GC, HPLC, MS, NMR, UV, IR, scintillation counters, P-32, animal facility, histopathology labs, isotope facilities, including P-32.
10. **Key personnel:**
    J. H. Weisburger
    D. Hoffmann
    D. Rose
    A. Rivenson
    K. Brunneman
    C. Meschter
    E. Fiala
11. **Staff:**
    Professional: 60, support staff: 160.
12. **GLP compliance:**
    Yes. Date of last inspection: August 5, 1988.
13. **FDA approval:**
14. **Accreditation:**
    AAALAC, AALAS.
15. **Communication capabilities:**
    Telephone number: (914) 592-2600
    Telefax number: (914) 592-6317
    Computer availability: VAX 750 and many microcomputers.
16. **Additional data:**
    Total Organic Halogens. International relationships, epidemiologic studies, smoking and health, nutrition and health. Editors of Preventive Medicine, Cell Biology and Toxicology. Editorial Boards of professional journals in areas of toxicology, medicine, pharmacology. Consultants to governmental agencies, professional societies, corporations.
17. **Key contact:**
    G. M. Williams, M.D.

1. **Name of company:**
   American Radiolabeled Chemicals, Inc.
2. **Address:**
   11612 Bowling Green Drive
   St. Louis, Missouri 63146
3. **Corporate officers:**
   Surendra Gupta, President
   Karen Gupta, Vice President
4. **Services offered:**
   Custom synthesis of radiolabeled chemicals.
5. **Tests performed:**
6. **Products tested:**
7. **Special expertise:**
   Radiolabeled chemical.
8. **Size of lab:**
   8,000 square feet.
9. **Equipment available:**
   Scintillation counter and scanner, electrophoresis, HPLC, Thin Layer Chromatography (TLC) scanner.
10. **Key personnel:**
    Phil Korb
    Chandranan Vora
    Witek Tomasik
    Paul Webb
11. **Staff:**
12. **GLP compliance:**
13. **FDA approval:**
    NRC inspection: License no. 24-21362-01. Date of last inspection: August 31, 1988.
14. **Accreditation:**
15. **Communication capabilities:**
    Telephone number: (314) 991-4545
    Telefax number: (314) 991-4692
    Telex number: 9102404101 AMMERADCHEM UQ
16. **Additional data:**
    Special services.
17. **Key contact:**
    Surendra Gupta

LISTING OF TOXICOLOGY LABORATORIES

1. **Name of company:**
   Amersham Corporation
2. **Address:**
   2636 S. Clearbrook Drive
   Arlington Heights, Illinois 60005
3. **Corporate officers:**
   T. S. Thompson, President
   D. G. Kolasinski, Vice President
   W. P. Fairbrother, Vice President
4. **Services offered:**
5. **Tests performed:**
6. **Products tested:**
7. **Special expertise:**
8. **Size of lab:**
9. **Equipment available:**
10. **Key personnel:**
11. **Staff:**
12. **GLP compliance:**
13. **FDA approval:**
14. **Accreditation:**
15. **Communication capabilities:**
    Telephone number: (312) 593-6300
    Telefax number: (312) 593-1044 or (312) 593-8236
16. **Additional data:**
    Special services.
17. **Key contact:**
    G. T. Anderson

1. **Name of company:**
   Ana-Lab Corporation
2. **Address:**
   2600 Dudley Road
   Kilgore, Texas 75662
3. **Corporate officers:**
   C. H. Whiteside, President
   D. L. Whiteside, Vice President
   J. A. Whiteside, Secretary/Treasurer
4. **Services offered:**
   Analytical chemistry, environmental services.
5. **Tests performed:**
   Chemical and biological analyses.
6. **Products tested:**
   Wastewater, solid waste, air, agricultural materials, foods.
7. **Special expertise:**
8. **Size of lab:**
   3,000 square feet.
9. **Equipment available:**
   GC, AA, GC/MS, UV-VIS, calorimetry, TOX, TOC, HPLC.
10. **Key personnel:**
    C. H. Whiteside, Ph.D., President/General Manager
    Bill Peery, Jr., Assisant Lab Manager
11. **Staff:**
    Chemists: 4, technicians: 20, other: 5.
12. **GLP compliance:**
13. **FDA approval:**
14. **Accreditation:**
    Oklahoma Water Board.
15. **Communication capabilities:**
    Telephone number: (214) 984-0551
    Computer availability: Yes.
16. **Additional data:**
    Analytical. Nontesting: (1) Expert testimony; (2) Operation and maintenance of water and wastewater systems.
17. **Key contact:**
    C. H. Whiteside, Ph.D.

LISTING OF TOXICOLOGY LABORATORIES

1. **Name of company:**
   Analytical Bio-Chemistry Laboratories, Inc.
2. **Address:**
   7200 East ABC Lane
   Columbia, Missouri 65202
3. **Corporate officers:**
   Ralph Waltz, President/CEO
   Charles Gehrke, Chairman of the Board
   David Stalling, Secretary
   Norman Rabjohn, Treasurer
4. **Services offered:**
   Total Organic Halogens. Analytical. Aquatic toxicology and bioconcentration, environmental fate studies, complete analytical support services, residue chemistry, residue studies in domestic animals, field studies, metabolism studies in rodents, domestic animals, and plants, metabolite isolation, characterization and identification, human clinical trials for pharmaceutical products, biopharmaceutical analyses, ecotoxicology, stability and dissolution studies.
5. **Tests performed:**
   Aquatic toxicology: acute, chronic tests and bioconcentration studies; environmental fate: degradation, metabolism, mobility studies; residue/metabolism: plants, domestic animals, rodents; field studies: terrestrial/aquatic dissipation; accumulation studies in crops and nontarget aquatic organisms.
6. **Products tested:**
   Agrichemicals, industrial chemicals, pharmaceuticals, effluents, cosmetics.
7. **Special expertise:**
   Regulatory consulting, metabolism chemistry, bioequivalency, bioavailability.
8. **Size of lab:**
   80,000 square feet.
9. **Equipment available:**
   GC, HPLC, MS, scintillation counter, 6500 watt xenon light source, environmental chambers, radiomatic radioactive TLC scanner, GPC, spectrophotometer, sample oxidizer, Beckman CLS main frame computer.
10. **Key personnel:**
    Lyle D. Johnson
    Carl Thompson
    Dr. Timothy Halls
    Dr. Michael Williams
    Michael Schofield
    Frank Selman
    Brenda Franklin
11. **Staff:**
    Total: 160, technicians: 39, PEs/scientists: 115, Ph.D./M.D.: 6.
12. **GLP compliance:**
    EPA audit. Date of last inspection: February 1989.
13. **FDA approval:**
    Nonclinical audit. Date of last inspection: September 1987.
14. **Accreditation:**
    AAALAC.
15. **Communication capabilities:**
    Telephone number: (314) 474-8579
    Telefax number: (314) 443-9033
    Telex number: 821814
16. **Additional data:**
    Total Organic Halogens. ABC Laboratories provides full-service contract research and analytical programs to assist clients with product registration and product safety studies. ABC's Instrument Division manufactures Model 1002B Automatic Gel Permeation Preparative Chromatograph, Model 601 Autovap(R)/GPC system, Model 602 Autovap(R) system.
17. **Key contact:**
    James B. Rabenold, Ph.D.

1. **Name of company:**
   Ani Lytics, Inc.
2. **Address:**
   360 Christopher Avenue
   Gaithersburg, Maryland 20879
3. **Corporate officers:**
   Saroj R. Das, Ph.D., President
   Walter F. Loeb, D.V.M., Ph.D., Vice President
4. **Services offered:**
   Hematology, pathology, and clinical chemistry testing; services in support of biomedical research, in vitro.
5. **Tests performed:**
   Clinical chemistry, hematology, pathology, coagulation, endocrinology, and other biomedical related clinical lab tests.
6. **Products tested:**
   Pharmaceuticals, biocompatible materials, blood, serum, tissue specimens.
7. **Special expertise:**
   Clinical chemistry, endocrinology, hematology, including methods development and esoteric tests. All methods species-appropriate.
8. **Size of lab:**
   1,500 square feet.
9. **Equipment available:**
   Random access chemistry analyzer, high-resolution spectrophotometer, hematology analyzers, cell analyzer, scintillation counter.
10. **Key personnel:**
    Walter F. Loeb, D.V.M., Ph.D.
    Saroj R. Das, Ph.D.
    Seena Polivy, B.S.
    Robert Heikkila, B.S.
11. **Staff:**
    Veterinary pathologist: 1, clinical chemical: 1, medical technologists: 2.
12. **GLP compliance:**
    Yes. Date of last inspection: April 1989.
13. **FDA approval:**
14. **Accreditation:**
15. **Communication capabilities:**
    Telephone number: (301) 921-0168
    Telefax number: (301) 977-0248
    Computer availability: IBM compatible.
16. **Additional data:**
    All methods used are based on Good Laboratory Practices (GLP) and appropriate for species tested; client-specified profiles, methods development and validation, esoteric testing and consultation to complement in-house testing, specimen collection specifications offered to optimize the quality of results, sex- and age-related statistical analyses of results.
17. **Key contact:**
    Saroj R. Das, Ph.D.

1. **Name of company:**
   Applied Genetics, Inc.
2. **Address:**
   1335 Gateway Dr. #2001
   Melbourne, Florida 32901-2619
3. **Corporate officers:**
   Dr. John C. Hozier, President
   Dr. Maria H. Lugo, Associate Director
4. **Services offered:**
   Research and development, cell characterization to biotechnology industry, molecular genetic studies.
5. **Tests performed:**
   Acute, in vitro.
6. **Products tested:**
   Pharmaceuticals.
7. **Special expertise:**
8. **Size of lab:**
   2,600 square feet.
9. **Equipment available:**
   Molecular genetic equipment.
10. **Key personnel:**
    Dr. John Hozier
    Dr. Maria Lugo
11. **Staff:**
    Technicians: 5, toxicologists: 2, scientists: 1.
12. **GLP compliance:**
    Practice good laboratory practices.
13. **FDA approval:**
14. **Accreditation:**
15. **Communication capabilities:**
    Telephone number: (407) 768-2048
    Telefax number: (407) 984-2890
16. **Additional data:**
    On-line with various data bases.
17. **Key contact:**
    Dr. Maria H. Lugo

1. **Name of company:**
   Argus Research Laboratories, Inc.
2. **Address:**
   935 Horsham Road
   Horsham, Pennsylvania 19044
3. **Corporate officers:**
   Mildred S. Christian, Ph.D., ATS President
   Alan M. Hoberman, Ph.D., Executive Vice President
4. **Services offered:**
   Reproductive and developmental toxicity testing, quality assurance monitoring and auditing.
5. **Tests performed:**
   Reproductive and development toxicity—in vivo and in vitro.
6. **Products tested:**
   Pharmaceuticals, chemicals, pesticides, cosmetics, medical devices materials.
7. **Special expertise:**
   Reproductive and developmental toxicology, quality assurance review, SOP development.
8. **Size of lab:**
   20,000 square feet.
9. **Equipment available:**
   All equipment required for conducting reproductive and developmental toxicity tests, including behavioral testing equipment and automated data collection and reporting.
10. **Key personnel:**
    Richard M. Hoar, Ph.D.
    E. Marshall Johnson, Ph.D.
    Elizabeth A. Lochry, Ph.D.
    Jane E. Goeke, Ph.D.
11. **Staff:**
    Senior level scientists: 10, technicians: B.S./M.S.: 15.
12. **GLP compliance:**
    EPA. Date of last inspection: November 1987. FDA. Date of last inspection: March 1988.
13. **FDA approval:**
    MAFF approval for Japan. Date of last inspection: 1987.
14. **Accreditation:**
15. **Communication capabilities:**
    Telephone number: (215) 443-8710
    Telefax number: (215) 443-8587
    Computer availability: Automated data collection for all work.
16. **Additional data:**
    Total Organic Halogens.
17. **Key contact:**
    Alan M. Hoberman, Ph.D.

1. **Name of company:**
   Battelle Memorial Institute
2. **Address:**
   505 King Avenue
   Columbus, Ohio 43201-2693
3. **Corporate officers:**
   Dr. Douglas E. Olesen, President
   Dr. William J. Madia, Senior Vice President
4. **Services offered:**
   Comprehensive services for approval/registration of drugs and chemicals, and for compliance with regulations protecting human health and safety and the environment.
5. **Tests performed:**
   Acute testing, subchronic testing, chronic toxicity and carcinogenicity evaluations, absorption, metabolism, distribution, and excretion studies, reproductive and developmental toxicology, cardiovascular and respiratory physiology/toxicology; wildlife and aquatic testing; herbicidal effects testing; herbicidal effects on target and nontarget plants; residue chemistry; environmental fate; product chemistry.
6. **Products tested:**
   Pesticides, air, water, soil, plants, sludge, mammalian toxicology, pharmacology.
7. **Special expertise:**
   Complete toxicology and pharmacology; clinical and comparative pathology; dose preparation laboratories; specialized lab facilities for performing neurobehavior, reproductive toxicology, and teratology studies in lab species.
8. **Size of lab:**
9. **Equipment available:**
   GC, GC/MS, high-resolution MS, MS/MS, FTIR, HPLC, NMR, ICP, AAS, IC.
10. **Key personnel:**
11. **Staff:**
12. **GLP compliance:**
    USDA. Date of last inspection: August 14, 1987.
13. **FDA approval:**
14. **Accreditation:**
    AAALAC.
15. **Communication capabilities:**
    Telephone number: (614) 424-5836
    Telefax number: (614) 424-5263
    Telex number: 24-5454
16. **Additional data:**
    Total Organic Halogens.
17. **Key contact:**
    Dr. Jake Halliday

1. **Name of company:**
   Bio/dynamics, Inc.
2. **Address:**
   Mettlers Road
   Box 2360
   East Millstone, New Jersey 08875-2360
3. **Corporate officers:**
   Geoffrey K. Hogan, Ph.D., DABT, President
   Aleksandar L. Knezevich, D.V.M., Senior Vice President/Director of Pathology
   Michael R. Norman, Senior Vice President/General Manager
   Ira W. Daly, Ph.D., DABT, Vice President/Director of Toxicology
4. **Services offered:**
   Preclinical toxicology/safety evaluation studies, metabolic and analytical chemistry, histotechnology and histopathological evaluations, laboratory facilities design and management, consulting and quality assurance services.
5. **Tests performed:**
   Pre-clinical in vivo toxicology studies of all types from acute through chronic, in all species via all routes of exposure.
6. **Products tested:**
   Pharmaceuticals, chemical, agricultural chemical, food, cosmetic, consumer products in U.S.A., Europe, and Japan.
7. **Special expertise:**
   Inhalation toxicology.
8. **Size of lab:**
   Facility: 185,000 square feet; animal accommodations: 104,000 square feet.
9. **Equipment available:**
   Equipment related to husbandry, pharmacy and feed, histopathological examinations, clinical examination, reproductive and teratogenic testing, inhalation testing, HPLC; GC; GC/MS; AA; metabolism instrumentation; general toxicology studies; electricity and pollution protecting equipment.
10. **Key personnel:**
    William J. Tierney, Ph.D., Director of Research and Market Development
    Paul E. Newton, Ph.D., Director of Inhalation Toxicology
    Joseph B. Townsend, Director of Quality Assurance
    Edward T. Greenstein, D.V.M.
    Carol S. Auletta, B.A., Associate Director of Toxicology
    Henry F. Bolte, D.V.M., Ph.D., Associate Director of Pathology
    John E. Atkinson, Ph.D., Manager of Rodent Toxicology
11. **Staff:**
    Large Animal Toxicology: 45, reproductive teratology: 13, inhalation toxicology: 20, chemistry: 18, pathology: 9, histology: 20.
12. **GLP compliance:**
    Yes. Date of last inspection: March 1987.
13. **FDA approval:**
    Yes. Date of last inspection: March 1987.
14. **Accreditation:**
    AAALAS.
15. **Communication capabilities:**
    Telephone number: (201) 873-2550
    Telefax number: (201) 873-3992
    Telex number: 844597 BIO DYN EMLS
16. **Additional data:**
    Conduct toxicity studies to meet or exceed worldwide GLP and regulatory compliance. Design, build, and/or manage toxicology laboratories; consulting and quality assurance services.
17. **Key contact:**
    Lynn J. Corrigan

1. **Name of company:**
   Bio-Life Associates Ltd.
2. **Address:**
   Route 3
   Box 156
   Neillsville, Wisconsin 54456
3. **Corporate officers:**
   Dale W. Fletcher
   Philomena K. Fletcher
4. **Services offered:**
   Toxicology testing; ecotoxicology testing.
5. **Tests performed:**
   Avian toxicology—LC50 and LD50—quail, duck, pheasant, passerines reproduction, residue and toxicity studies, dermal toxicity studies, acute and chronic neurotoxicity; terrestrial field toxicology—level I and II studies; large animal toxicology—milk and meat residue studies, radiolabeled studies; special animal toxicology; residue monitoring—crop, tissue, diets.
6. **Products tested:**
   Pesticides, cosmetics, pharmaceuticals, industrial chemicals.
7. **Special expertise:**
   Safety evaluations.
8. **Size of lab:**
   15,140 square feet.
9. **Equipment available:**
   Animal toxicology equipment.
10. **Key personnel:**
    D. W. Fletcher, B.S.
    D. E. Gordon, Ph.D., D.V.M.
    G. M. Rand, Ph.D.
    C. A. Pedersen, B.S.
    M. I. Leonard, Ph.D.
    B. J. Cooksey, Ph.D.
    R. W. Fish, D.V.M.
11. **Staff:**
    Toxicologists: 4, veterinarians: 2, technicians: 12.
12. **GLP compliance:**
    EPA/FDA. Date of last inspection: December 3, 1986.
13. **FDA approval:**
    Yes. Date of last inspection: in conjunction with EPA inspection on December 3, 1986.
14. **Accreditation:**
    Not available for avian facility.
15. **Communication capabilities:**
    Telephone number: (715) 743-4557
    Computer availability: CPT 9000 with modem for on-line capability.
16. **Additional data:**
    Total Organic Halogens.
17. **Key contact:**
    Dale W. Fletcher

1. **Name of company:**
   Biological Test Center
2. **Address:**
   2525 McGaw Avenue
   Irvine, California 92714
3. **Corporate officers:**
4. **Services offered:**
   Toxicology, pharmacokinetics, absorption, distribution, metabolism, excretion (ADME) studies, analytical testing, QA/QC testing.
5. **Tests performed:**
   In vivo: acute, subchronic, pharmacokinetic, ADME, ophthalmic studies on all species of laboratory animals. In vitro: cytotoxicity, hemolysis, and LAL testing.
6. **Products tested:**
   Pharmaceuticals, medical devices, chemicals, pesticides, cosmetics.
7. **Special expertise:**
   Pharmacokinetics, ADME studies, toxicology, biocompatibility, QA/QC testing.
8. **Size of lab:**
   32,000 square feet.
9. **Equipment available:**
   Microsurgery equipment, scintillation and gamma counters, GC, HPLC, sample oxidizers.
10. **Key personnel:**
11. **Staff:**
    Supervisors: 9, technicians: 32.
12. **GLP compliance:**
    Yes. Date of last inspection: August 18, 1988.
13. **FDA approval:**
    Yes. Date of last inspection: August 18, 1988.
14. **Accreditation:**
    AAALAC, AALAS.
15. **Communication capabilities:**
    Telephone number: (714) 660-3185
    Telefax number: (714) 660-2565
    Telex number: 4970362
16. **Additional data:**
    Total Organic Halogens.
17. **Key contact:**
    Paul Mazur, Ph.D.

LISTING OF TOXICOLOGY LABORATORIES

1. **Name of company:**
   BIOMED, Inc.
2. **Address:**
   1720 130th Ave., N.E.
   Bellevue, Washington 98005-2203
3. **Corporate officers:**
   Joseph Ashley, Chairman/CEO
   Mark Levine, President
   Sally Majnarich, Corporate Treasurer
4. **Services offered:**
   Analytical and microbiological capability including food, feed, water, wastewater, solid, hazardous waste using EPA methods. Chronic and acute bioassays using rodents, fish, and marine organisms.
5. **Tests performed:**
   Acute oral toxicity, dermal toxicity, acute inhalation, primary eye irritation, primary dermal irritation, dermal sensitization; subchronic testing; chronic studies; mutagenicity—Ames testing; metabolism studies. Laboratory has in-house capabilities for acute and subacute chronic bioassays using rodents, fish, and/or marine organisms.
6. **Products tested:**
   Cosmetics, foods, feed, water, wastewater, solid, hazardous waste, pharmaceuticals.
7. **Special expertise:**
   Bioassays using rodents, fish, or marine organisms.
8. **Size of lab:**
   20,000 square feet.
9. **Equipment available:**
   GC, HPLC.
10. **Key personnel:**
    Craig E. Delphy, Ph.D., Director of Laboratory Services
    John J. Majnarich, Ph.D., Scientific Director
    Thomas D. Goodrich, Ph.D., Manager of Research and Development
    Warren C. Ladiges, D.V.M., ACLAM
11. **Staff:**
    Toxicologists: 2, technicians: 3.
12. **GLP compliance:**
    Yes.
13. **FDA approval:**
    FDA GLP: Yes.
14. **Accreditation:**
15. **Communication capabilities:**
    Telephone number: (206) 882-0448
    Telefax number: (206) 882-2678
    Telex number: 283803 BIOM
16. **Additional data:**
    Total Organic Halogens. Custom fermentation.
17. **Key contact:**
    Mark Levine

1. **Name of company:**
   Bionetics Research, Inc.
2. **Address:**
   5516 Nicholson Lane
   Kensington, Maryland 20895-1078
3. **Corporate officers:**
   Lloyd Moores, President
   Mangalas Sarngadharan, Vice President of Scientific Affairs
   Gary Mills, General Counsel
4. **Services offered:**
   Evaluation of disinfectants against HIV, isolation of HIV, evaluation of drugs against HIV, evaluation of blood products for safety, custom-designed studies relating to HIV.
5. **Tests performed:**
   In vitro systems employing tissue culture cells.
6. **Products tested:**
   Disinfectants, blood, body fluids or tissues, drugs, blood products.
7. **Special expertise:**
   Infectivity and virus isolation of the human immunodeficiency (AIDS) virus.
8. **Size of lab:**
   1,800 square feet.
9. **Equipment available:**
   Centrifuges, scintillation counter, bio-hazard safety hoods, special safety centrifuges for containment, ultra-low temperature storage facilities.
10. **Key personnel:**
    P. D. Markham, Ph.D., Scientific Director
    S. C. Tondreau, Study Director
11. **Staff:**
    Biologists: 5.
12. **GLP compliance:**
    Yes. Date of last inspection: August 3, 1987.
13. **FDA approval:**
    EPA. Date of last inspection: August 4, 1987 for protocols for virus testing services.
14. **Accreditation:**
    AAALAC, USDA, OPRR.
15. **Communication capabilities:**
    Telephone number: (301) 881-5600
    Telefax number: (301) 984-3608
    Telex number: 89-8369
16. **Additional data:**
    Total Organic Halogens.
17. **Key contact:**
    Sue C. Tondreau

1. **Name of company:**
   Biospherics, Inc.
2. **Address:**
   12051 Indian Creek Court
   Beltsville, Maryland 12051
3. **Corporate officers:**
   Gilbert V. Levin, President
   M. Karen Levin, Vice President
   Mary Jo Szorady, Vice President
   Robert G. Edwards, Ph.D., Vice President of Science Services
   John W. Kraus, Vice President of Finance
4. **Services offered:**
   Analytical, product testing, environmental studies, groundwater contaminant assessment, industrial hygiene, health and safety plans, asbestos programs.
5. **Tests performed:**
   Full-service capability in organic and inorganic analytical chemistry and microbiology, leach rates, method development/validation, laboratory studies for registration of agricultural/industrial chemicals and veterinary products under EPA/FIFRA or TSCA and FDA regulations.
6. **Products tested:**
   Soil, water, air, sludge, sediment, crops, animal tissue, feeds, foods, drummed wastes, asbestos insulation.
7. **Special expertise:**
   Environmental sample analysis, residue sample analysis, waste or bulk characterization, industrial hygiene sample analysis.
8. **Size of lab:**
   30,000 square feet.
9. **Equipment available:**
   Air sampling pumps, air Hi Vol biological samplers, explosimeters, HIAC particulate analyzers, IC, IR, UV-VIS, TOC, TOX, GC, GC/MS, HPLC, ICP, AA.
10. **Key personnel:**
    Nancy Cargile, Ph.D., Director, Laboratory Division
    John Kearns, Marketing Director, Laboratory Division
    Marcella Saynuk, Assistant Director, Laboratory Division
    Mark Gudnason, Manager, Laboratory Operations
    Stuart Z. Cohen, Ph.D., Manager, Groundwater Department
    Len Burelli, Director, Industrial Hygiene Division
11. **Staff:**
    Industrial hygienists: 25, ecologists: 3, oceanographer: 1, chemists: 25, geologists: 2, sanitary engineers: 2, chemical technicians: 6, mechanical engineer: 1, surveyors: 6.
12. **GLP compliance:**
    Yes. Date of last inspection: May 19–22, 1986.
13. **FDA approval:**
    Yes. Date of last inspection: August 31, 1987.
14. **Accreditation:**
    Certified by Virginia for water analysis; by Maryland in drinking water analysis; by New Jersey for water analysis.
15. **Communication capabilities:**
    Telephone number: (301) 369-3900
    Telefax number: (301) 725-4908
    Telex number: 898 072
    Computer availability: PC network.
16. **Additional data:**
    Total Organic Halogens.
17. **Key contact:**
    Robert G. Edwards, Ph.D.

1. **Name of company:**
   Bushy Run Research Center
2. **Address:**
   R.D. 4, Mellon Rd.
   Export, Pennsylvania 15632
3. **Corporate officers:**
   Fred R. Frank, Director
   William M. Snellings, Associate Director
   Edward H. Fowler, Diplomate ACVP/Associate Director
4. **Services offered:**
   Toxicology testing; gross, anatomic, and clinical pathology; analytical services.
5. **Tests performed:**
   Acute, short-term, subchronic, chronic; neurotoxicity testing; genotoxicity screening; material balance and toxicokinetic studies; cytotoxicity screening.
6. **Products tested:**
   Industrial chemicals, pesticides, cosmetics, pharmaceuticals.
7. **Special expertise:**
   Developmental toxicology, inhalation toxicology, neurotoxicology, pulmonary irritation testing, pharmacokinetics—in vitro skin penetration.
8. **Size of lab:**
   60,000 square feet.
9. **Equipment available:**
   Scanning and transmission electron microscopes; energy dispersive x-ray. Almost all analytical equipment necessary for test chemical concentration analysis, pharmacokinetic determination, and metabolic identification.
10. **Key personnel:**
    Ronald S. Slesinski, Ph.D., DABT/Assistant Director
    Rochelle W. Tyl, Ph.D., DABT/Assistant Director
    Fred R. Frank, Ph.D., Director
    Edward H. Fowler, DVM, Ph.D., Diplomate ACVP/Associate Director
    William M. Snellings, Ph.D./Associate Director
11. **Staff:**
12. **GLP compliance:**
    Yes. Date of last inspection: August 23–26, 1988.
13. **FDA approval:**
    Yes. Date of last inspection: September 13, 1983.
14. **Accreditation:**
    AAALAC, institutional member of AALAS, accredited by TLA board.
15. **Communication capabilities:**
    Telephone number: (412) 733-5200
    Telefax number: (412) 733-4804
16. **Additional data:**
    Total Organic Halogens. As part of the Mellon Institute, services can be expanded to include solid aerosol material characterization.
17. **Key contact:**
    William M. Snellings, Ph.D./Associate Director

LISTING OF TOXICOLOGY LABORATORIES

1. **Name of company:**
   Chemsyn Science Laboratories, Member of the Specialty Materials Division, Eagle-Picher Industries, Inc.
2. **Address:**
   13605 W. 96th Terrace
   Lenexa, Kansas 66215
3. **Corporate officers:**
   Thomas K. Dobbs, Vice President/Department Manager
4. **Services offered:**
   Custom synthesis of a wide variety of organic compounds, including mass and radiolabeled compounds and bulk pharmaceuticals; custom analytical services, including analyses by HPLC, GC, MS, and NMR.
5. **Tests performed:**
   Chemical analysis and characterization by NMR, MS, HPLC, GC. Analysis of radiolabeled compounds by radio-HPLC and TLC.
6. **Products tested:**
7. **Special expertise:**
   Radiolabeled synthesis involving 3H and 14C, synthesis of carcinogenic or toxic materials, production of bulk pharmaceuticals.
8. **Size of lab:**
   8,000 square feet.
9. **Equipment available:**
   GC, NMR, scintillation counters, HPLC (analytical and preparatory), radiochromatogram scanners, UV/VIS spectrophotometer.
10. **Key personnel:**
    Jim Wiley, Technical Manager
    Dr. Robert Roth, Synthesis Group Leader
    Dr. Paul Ruehl, Synthesis Group Leader
    Tom Spencer, Synthesis Group Leader
    Beth Minter, Analytical Services Group Leader
11. **Staff:**
    Technical: 30.
12. **GLP compliance:**
    GMP/GLP.
13. **FDA approval:**
    Drug Master File established at the FDA. Note: Licensed by Kansas and NRC to handle tritium and carbon-14.
14. **Accreditation:**
15. **Communication capabilities:**
    Telephone number: (913) 541-0525, (800) 233-6643
    Telefax number: (913) 888-3582
    Telex number: 910-840-3270
16. **Additional data:**
    Special services.
17. **Key contact:**
    Lisa Bosch

1. **Name of company:**
   Colorado Histo-Prep., Inc.
2. **Address:**
   Box 8644
   Fort Collins, Colorado 80524
3. **Corporate officers:**
   Janet E. Maass, President
   M. H. Maass, Vice President
   Susan M. Maass, Secretary
4. **Services offered:**
   Histology preparation of microscopic slides.
5. **Tests performed:**
   Process all testing animal types, including exotic animals.
6. **Products tested:**
7. **Special expertise:**
   Immunohistochemistry, plastics, and special stains.
8. **Size of lab:**
   1,500 square feet.
9. **Equipment available:**
   Tissue processors, microtomes, embedding centers, microwave, computer, axillary equipment.
10. **Key personnel:**
    Janet Maass, Histotechnologist, Cytotoechnologist
11. **Staff:**
    Histology technicians: 5, histology assistants: 4.
12. **GLP compliance:**
    Every study is audited by the quality assurance unit. This amounts to weekly inspections.
13. **FDA approval:**
    Yes. Date of last inspection: January 1987, with no FDA 483 issued.
14. **Accreditation:**
15. **Communication capabilities:**
    Telephone number: (303) 493-2660
16. **Additional data:**
17. **Key contact:**
    Janet E. Maass

1. **Name of company:**
   Comparative Toxicology Laboratories, VCS: Kansas State University
2. **Address:**
   Manhattan, Kansas 66506
3. **Corporate officers:**
   Frederick W. Oehme, Director
4. **Services offered:**
   Analytical and forensic testing, research for chemical safety, investigations of mechanisms of action and effects of chemicals.
5. **Tests performed:**
   All tests except for carcinogenicity studies.
6. **Products tested:**
   All products may be tested, but naturally occurring compounds, new drugs and chemicals, and environmental compounds are commonly studied.
7. **Special expertise:**
   Comparative toxicity studies using a wide variety of animals; metabolic and tracer studies of excretion and residues.
8. **Size of lab:**
   6,000 square feet plus holding areas for animals.
9. **Equipment available:**
   All analytical instrumentation needed for biotransformation studies and tracer research. Animal facilities for large and small research animals.
10. **Key personnel:**
    F. W. Oehme, Professor and Director
11. **Staff:**
    Toxicologists: 3, technicians: 2, research assistants: 8.
12. **GLP compliance:**
    Yes. Date of last inspection: 1982.
13. **FDA approval:**
14. **Accreditation:**
    AAALAC.
15. **Communication capabilities:**
    Telephone number: (913) 532-5679
    Computer availability: Yes.
16. **Additional data:**
    Total Organic Halogens. This is a university research and testing laboratory.
17. **Key contact:**
    F. W. Oehme

1. **Name of company:**
   Cosmopolitan Safety Evaluation, Inc.
2. **Address:**
   Statesville Quarry Rd.
   Box 71
   Lafayette, New Jersey 07848
3. **Corporate officers:**
   Geoffrey R. Robbins, M.R.C.V.S., DABT, CEO
4. **Services offered:**
   Toxicology.
5. **Tests performed:**
   Acute, in vitro, rats, mice, rabbits, guinea pig, hamsters.
6. **Products tested:**
   Pesticides, cosmetics, pharmaceuticals, industrial chemicals, microbiological.
7. **Special expertise:**
   Regulatory.
8. **Size of lab:**
   20,000 square feet.
9. **Equipment available:**
   Standard.
10. **Key personnel:**
    Gerry Rosenfeld, Laboratory Manager
    Vanessa White, Quality Assurance Unit
11. **Staff:**
    Toxicologists: 3, technicians: 6.
12. **GLP compliance:**
    Yes. Date of last inspection: March 1989.
13. **FDA approval:**
    Yes. Date of last inspection: March 1989.
14. **Accreditation:**
15. **Communication capabilities:**
    Telephone number: (201) 383-6253
    Telefax number: Yes, restricted to clients
    Computer availability: Hayes compatible modem.
16. **Additional data:**
    Consultation by American Board of Toxicology diplomats (2) and expert witnesses (3)—pesticide/herbicide and oncogenicity.
17. **Key contact:**
    Dr. G. R. Robbins

LISTING OF TOXICOLOGY LABORATORIES

1. **Name of company:**
   Dawson Research Corporation
2. **Address:**
   Box 620666
   Orlando, Florida 32862-0666
3. **Corporate officers:**
   Thomas E. Murchison, D.V.M., M.Sc., Ph.D., President/Scientific Director
4. **Services offered:**
   Preclinical safety testing, toxicology, pathology, clinical pathology, histology, histopathology.
5. **Tests performed:**
   Acute; subacute; chronic toxicity; carcinogenicity; teratology on mice, rats, rabbits, guinea pigs, dogs, ferrets, nonhuman primates.
6. **Products tested:**
   Pesticides, cosmetics, pharmaceuticals, chemicals.
7. **Special expertise:**
   Pathology, ophthalmology, cardiology, neurology.
8. **Size of lab:**
   26,000 square feet.
9. **Equipment available:**
   Gemini autoanalyzer, HPLC, electrophoresis, EKG, biomicroscope.
10. **Key personnel:**
    Gordon T. Geisler
    Elizabeth S. Levesque
11. **Staff:**
    Pathologist: 1, quality assurance personnel: 2, reporting personnel: 3, toxicologist: 1, medical technologist: 1, animal caregiver: 1, consultant scientists: 5, technicians: 9.
12. **GLP compliance:**
    Yes. Date of last inspection: December 1986.
13. **FDA approval:**
    Yes. Date of last inspection: December 1986.
14. **Accreditation:**
    AALAS, USDA, FDA, EPA.
15. **Communication capabilities:**
    Telephone number: (407) 851-3110
    Telefax number: (407) 851-3110
    Computer availability: IBM PS/2, Model 2.
16. **Additional data:**
    Total Organic Halogens. Expert witness, safety evaluation, consultation, medical-legal transcription, library research.
17. **Key contact:**
    Thomas E. Murchison, D.V.M., Ph.D.

1. **Name of company:**
   Ecology & Environment, Inc., Analytical Services Center
2. **Address:**
   4285 Genesee Street
   Buffalo, New York 14225
3. **Corporate officers:**
   Gerhard J. Neumaier, President
   Frank B. Silvestro, Executive Vice President
   Gerald A. Strobel, P.E., Executive Vice President
   Ronald L. Frank, Executive Vice President
   Roger Gray, Senior Vice President
   Gerard A. Gallagher, Jr., Senior Vice President
   Eugene R. Mruk, AICP, Vice President
   Laurence Brickman, Ph.D., Vice President
4. **Services offered:**
   Sampling of contaminated soil/waste, wastewater, groundwater, surfacewater, air emissions; analysis of air, water, and contaminated soil/waste samples; aquatic toxicology; asbestos testing.
5. **Tests performed:**
   Acid fraction, aquatic toxicity, asbestos, base-neutral fraction, corrosivity, extraction procedures for oily wastes and toxicity, ignitability, metals, paint filter test, pesticides, phenolics, polychlorinated biphenyls.
6. **Products tested:**
   All environmental samples.
7. **Special expertise:**
   GC/MS, electron microscopy.
8. **Size of lab:**
   16,400 square feet.
9. **Equipment available:**
   GC/MS, Tekmar LSC-2 liquid sample concentrators for volatile organic analysis, Tekmar Model 4200 automatic heated sampler module for volatile soil analysis, IC, GC, Perkin Elmer Plasma II inductively coupled argon plasma spectrophotometer, AAS (Atomic Absorption Spectrophotometer), Microscope equipped for bulk asbestos analysis.
10. **Key personnel:**
    Andrew Clifton, Director of Analytical Services
    Gary Hahn, Laboratory Manager
11. **Staff:**
    Total: 50.
12. **GLP compliance:**
    N/A.
13. **FDA approval:**
    N/A.
14. **Accreditation:**
15. **Communication capabilities:**
    Telephone number: (716) 631-0360
16. **Additional data:**
    Analytical. Provides a complete range of environmental scientific and engineering consulting services—hazardous waste management, spill emergency response, asbestos removal management, underground storage tank management, environmental auditing and impact analysis, hazards and risk analysis, archaeological surveys, hydrogeological studies, industrial hygiene studies, site remedial engineering, environmental engineering. Ecology & Environment, Inc., is headquartered in Buffalo, NY, and has offices in Albany, Philadelphia, Washington, Tallahassee, Oak Ridge, Chicago, St. Louis, Kansas City, Baton Rouge, Houston, Dallas, Denver, Los Angeles, San Francisco, Seattle, and Anchorage. Represented in 21 foreign countries. Total employment is approximately 1,000.
17. **Key contact:**
    John Gartner

LISTING OF TOXICOLOGY LABORATORIES

1. **Name of company:**
   ECS/Normandeau
2. **Address:**
   Box 1393
   Aiken, South Carolina 29802
3. **Corporate officers:**
   Gene Eidson, Vice President/General Manager
   James O'Hara, Vice President
   Henry Kania, Vice President
4. **Services offered:**
   Acute and chronic freshwater and marine toxicity testing, toxicity identification evaluations, toxicity reduction evaluations, sediment toxicity tests, soil and sludge toxicity tests.
5. **Tests performed:**
   Chemical analysis: organic/inorganic analyses for waste water, drinking water, and hazardous wastes.
6. **Products tested:**
   Industrial/municipal/environmental water; sludge, sediments, dredged material; hazardous wastes.
7. **Special expertise:**
   Performance of freshwater and marine toxicity tests, nutritional requirements of freshwater zooplankton, TIG/TRE studies.
8. **Size of lab:**
   Aquatic Toxicology Lab: 2,000 square feet; Chemical and biology: 10,000 square feet.
9. **Equipment available:**
   GC, GC/MS, ICAP, Furnace AA, Flame AA, Technicon Traacs System.
10. **Key personnel:**
    Kathleen E. Trapp, Ph.D.
    Eric Korthals, M.S.
11. **Staff:**
12. **GLP compliance:**
    N/A.
13. **FDA approval:**
    N/A.
14. **Accreditation:**
    NC, SC, GA certified; have applied to FL, NH, CT, NJ for certification.
15. **Communication capabilities:**
    Telephone number: (803) 652-2206
    Telefax number: (803) 652-7428
16. **Additional data:**
    Analytical. Extensive ecological capabilities (fisheries, limnology, etc.), complete chemical capabilities, expert witness, research level projects requiring expert testimony, marine and freshwater chemistry, wetlands assessment.
17. **Key contact:**
    Kathleen E. Trapp, Ph.D.

1. **Name of company:**
   Education & Research Foundation, Inc.
2. **Address:**
   2602 Langhorne Road
   Lynchburg, Virginia 24501
3. **Corporate officers:**
   Claire G. Whitmore, President
   Bert Mathews, Vice President/Director
4. **Services offered:**
   Dermatology efficacy studies, patch, photopatch, sunscreen, antiperspirant testing of male and female panels, deodorant efficacy, ethical drug studies, gastroenterological safety and efficacy studies.
5. **Tests performed:**
   Physical, human clinical, safety, and efficacy.
6. **Products tested:**
   Cosmetics, pharmaceuticals.
7. **Special expertise:**
   SPF, dermal safety, dandruff, acne, IND—phases I, II, III.
8. **Size of lab:**
   6,000 square feet.
9. **Equipment available:**
   Zenon lamp.
10. **Key personnel:**
    Janet Hickman, M.D.; Medical Director
11. **Staff:**
    Total: 12, part-time technicians: 30.
12. **GLP compliance:**
13. **FDA approval:**
14. **Accreditation:**
    American Academy of Dermatology, SCC.
15. **Communication capabilities:**
    Telephone number: (804) 847-5695
    Telefax number: (804) 846-1707
16. **Additional data:**
    Total Organic Halogens. All protocols submitted to Institutional Review Board.
17. **Key contact:**
    Bert Mathews, Vice President/Director

1. **Name of company:**
   Enseco, Inc.
2. **Address:**
   Doaks Lane at Little Harbor
   Marblehead, Massachusetts 01945
3. **Corporate officers:**
   William D. Rucklhaus, Chairman of the Board
   Harvey G. Felsen, President
4. **Services offered:**
   Aquatic toxicology, analytical chemistry.
5. **Tests performed:**
   Acute, chronic, partial-chronic, bioconcentration with 40 species of freshwater and marine organisms.
6. **Products tested:**
   Pesticides, toxic substances, industrial chemicals, wastes, effluents.
7. **Special expertise:**
8. **Size of lab:**
   Aquatic toxicology facility: 5,000 square feet; analytical chemistry facility: 200,000 square feet.
9. **Equipment available:**
   GC/MS, AA, HPLC, ICP.
10. **Key personnel:**
11. **Staff:**
    Biologists: 20, chemists and support staff: more than 500.
12. **GLP compliance:**
    Yes. Date of last inspection: April 1988.
13. **FDA approval:**
    N/A.
14. **Accreditation:**
15. **Communication capabilities:**
    Telephone number: (617) 639-2695
    Telefax number: (617) 639-2637
    Computer availability: Yes.
16. **Additional data:**
    Total Organic Halogens.
17. **Key contact:**
    Timothy J. Ward

1. **Name of company:**
   Enviroscan, Inc.
2. **Address:**
   303 W. Military Road
   Rothschild, Wisconsin 54474
3. **Corporate officers:**
   Alfred Slatin, CEO
   William M. Copa, President
   James W. Barr, Vice President
   Douglas G. Schubring, Vice President
   James M. Force, Vice President
4. **Services offered:**
   Testing and monitoring ground water, process streams, wastes and waste streams, sludges, soils, hazardous wastes.
5. **Tests performed:**
   All chemical analyses associated with water, wastewater, soils, and sludges.
6. **Products tested:**
   Organics, inorganics, water, wastewater, soils, sludges.
7. **Special expertise:**
   Environmental analysis, priority pollutants, soils, groundwater.
8. **Size of lab:**
   6,000 square feet.
9. **Equipment available:**
   GC/MS, GC, HPLC, AA, ICP, Technicon AutoAnalyzer II, IR, IC, UV-VIS, total organic carbon analyzer, calorimeter, RCRA testing, microscopes, Purge and Trap.
10. **Key personnel:**
    James R. Salkowski, B.S., M.E.A.S., Manager of Inorganic Laboratory
    David L. Schumacher, B.S., Manager of Organic Laboratory
11. **Staff:**
    Chemists: 7, field specialists: 2, technicians: 10, support: 3.
12. **GLP compliance:**
    Adhere to GLP practices.
13. **FDA approval:**
14. **Accreditation:**
    Certified by WI DNR, WI Environmental Laboratory Association.
15. **Communication capabilities:**
    Telephone number: (715) 359-7226, (800) 338-7226
    Telefax number: (715) 355-3219
    Telex number: 29-0495
    Computer availability: DEC 20/70.
16. **Additional data:**
    Analytical. Real estate site investigations, field sampling, regulatory report assistance.
17. **Key contact:**
    James M. Force

1. **Name of company:**
   Essex Testing
2. **Address:**
   799 Bloomfield Avenue
   Verona, New Jersey 07044
3. **Corporate officers:**
   Harold Schwartz, Ph.D., President
   Robert W. Shanahan, Ph.D., Vice President/Technical Director
4. **Services offered:**
   Human clinical trials—safety and tolerance; studies on drugs, cosmetics, toiletries, specialty chemicals as adhesives.
5. **Tests performed:**
   Skin patch testing for irritation/sensitization, drug tolerance studies, clinical trials—bioavailability.
6. **Products tested:**
   Pharmaceuticals (over the counter and ethical, cosmetics, toiletries, specialty chemicals).
7. **Special expertise:**
   Safety and efficacy human clinical trials, especially dermal testing; gynecological and ophthalmological expertise.
8. **Size of lab:**
   5,000 square feet.
9. **Equipment available:**
   Solar simulators, UVA emitters, slit lamp, NMR, EKG.
10. **Key personnel:**
    Michael Frentzko, B.S., Laboratory Manager
    Hyman Menduke, Ph.D., Consultant Statistician
11. **Staff:**
    Technicians: 18, statistician: 1.
12. **GLP compliance:**
    N/A.
13. **FDA approval:**
    N/A.
14. **Accreditation:**
    AAALAC.
15. **Communication capabilities:**
    Telephone number: (201) 857-9541
    Telefax number: (201) 857-9662
    Computer availability: Yes.
16. **Additional data:**
    Total Organic Halogens. Essex Testing does not perform nonclinical (animal) studies.
17. **Key contact:**
    Harold Schwartz

1. **Name of company:**
   Food and Drug Research Laboratories, Division of Enviro/Analysis Corporation
2. **Address:**
   Box 107, Rt. 17C
   Waverly, New York 14892
3. **Corporate officers:**
   E. Corbin McGriff, Jr., President/CEO of Enviro/Analysis
   Roger Eaglen, Chairman of the Board
   John A. Biesemeier, President/CEO of Food and Drug Research Laboratories
   James Laveglia, Vice President of Food and Drug Research Laboratories
4. **Services offered:**
   Toxicology testing, analytical chemistry services, archiving services.
5. **Tests performed:**
   Acute, subchronic, chronic, sensitization.
6. **Products tested:**
   Pesticides, industrial chemicals, cosmetics, food additives, pharmaceutical.
7. **Special expertise:**
   Inhalation toxicology.
8. **Size of lab:**
   23,000 square feet.
9. **Equipment available:**
   GLC, HPLC, GC/MS, ICP, AA.
10. **Key personnel:**
    John Biesemeier, B.S., President, Study Director
    James Laveglia, Ph.D., Vice President, Director of Toxicology, Study Director
    Julie Moon, B.S., Study Director
    Fred Paul, B.S., Manager Quality Assurance and Archives
    Beth Reagan, AAS, Study Director
11. **Staff:**
    Managerial: 7, technical: 10.
12. **GLP compliance:**
    Yes. Date of last inspection: March 1987, joint inspection with EPA.
13. **FDA approval:**
    Yes. Date of last inspection: February 1988.
14. **Accreditation:**
    AAALAC, NIH.
15. **Communication capabilities:**
    Telephone number: (607) 565-8131
    Telefax number: (607) 565-7420
16. **Additional data:**
    Total Organic Halogens.
17. **Key contact:**
    John A. Biesemeier

LISTING OF TOXICOLOGY LABORATORIES

1. **Name of company:**
   Harris Laboratories, Inc.
2. **Address:**
   624 Peach Street
   Box 80837
   Lincoln, Nebraska 68501
3. **Corporate officers:**
   James McClurg, Ph.D., President, Harris Laboratories, Life Sciences
   Ronald L. Harris, President/CEO, Harris Technology Group
   Robert B. Harris, Chairman of the Board, Harris Technology Group
4. **Services offered:**
   Clinical pharmaceutical research—phases I, II, III, IV; bioanalytical services; dermatological and consumer use studies.
5. **Tests performed:**
   Clinical: bioavailability, metabolism, radiolabeled, rising dose tolerance, patient trials; agricultural: soil nutrient analysis, pesticide/herbicide residue, water analysis, animal feed analysis.
6. **Products tested:**
   Pharmaceuticals, consumer products, sun protection products, soil, water.
7. **Special expertise:**
   Clinical pharmaceutical studies, consumer use studies, dermatological testing.
8. **Size of lab:**
   100,000 square feet.
9. **Equipment available:**
   GC, HPLC, RIA, AA, ICP, GC/MS, microbiology.
10. **Key personnel:**
    James C. Kisicki, M.D.
    Charles Ryan, Ph.D.
    James D. Hulse, Ph.D.
    Gene Heath, Ph.D.
    Oliver Wzite, D.D.S.
    David Anderson
    Tom Buelt
11. **Staff:**
    Professional—chemists, healthcare personnel, agronomists: 250, support: 150.
12. **GLP compliance:**
    Yes.
13. **FDA approval:**
    Several in past 3 years.
14. **Accreditation:**
    AALA, CAP (pathology, clinical lab).
15. **Communication capabilities:**
    Telephone number: (402) 476-2811
    Telefax number: (402) 476-7598
    Telex number: 230-469-018
    Computer availability: (402) 475-1770.
16. **Additional data:**
    Total Organic Halogens.
17. **Key contact:**
    Robert P. Marshall

1. **Name of company:**
   Hazards Research Corporation
2. **Address:**
   200 East Main Street
   Rockaway, New Jersey 07866
3. **Corporate officers:**
   Dr. Chester Grelecki, President
   William J. Cruice, Vice President/Treasurer
   Harry N. McClary, Secretary
4. **Services offered:**
   Consulting firm. No laboratory—provide lab services by use of other companies' lab facilities, using their personnel or HRC personnel, as appropriate.
5. **Tests performed:**
   Fire and explosion studies.
6. **Products tested:**
   All chemicals and related materials, including foods, fuels, agricultural chemicals, industrial chemicals, explosives, drugs, metals, plastics.
7. **Special expertise:**
   Fire and explosion studies.
8. **Size of lab:**
9. **Equipment available:**
10. **Key personnel:**
    Dr. Chester Grelecki, Chief Scientist
    William J. Cruice, Vice President, Associate Chief Scientist
    George J. Petino, Jr., Chief Mechanical Engineer
    Steven J. Tunkel, Chief Chemical Engineer
    Francis X. Bender, Chemical Engineer
    Harry N. McClary, Technical Specialist
11. **Staff:**
    Physical chemists: 2, mechanical engineer: 1, chemical engineers: 2, technical specialist: 1.
12. **GLP compliance:**
    N/A.
13. **FDA approval:**
    N/A.
14. **Accreditation:**
    N/A.
15. **Communication capabilities:**
    Telephone number: (201) 627-4560
    Telefax number: (201) 627-0015
16. **Additional data:**
    Analytical.
17. **Key contact:**
    William J. Cruice

LISTING OF TOXICOLOGY LABORATORIES

1. **Name of company:**
   Hazleton Laboratories America, Inc.
2. **Address:**
   3301 Kinsman Boulevard
   Madison, Wisconsin 53704
3. **Corporate officers:**
   Donald P. Nielsen   Carl C. Schwan
   Roy M. Dagnall   Robert E. Conway
4. **Services offered:**
   Toxicology, metabolism, pharmacokinetics, clinical evaluation, drug safety and efficacy, analytical services and support, nutritional analysis, pesticide and residue analysis, environmental fate.
5. **Tests performed:**
   Metabolism, dermal absorption and sensitization, neurotoxicity, rodent and dog feeding studies, oncogenicity, teratogenicity, mutagenicity, carcinogenicity; nature of residue in plants and animals, preclinical toxicology, acute, subchronic, chronic, chemistry, clinical, nutritional.
6. **Products tested:**
   Food and feed, household and personal care products, pharmaceuticals, pesticides, animal health products, agricultural and industrial chemicals, medical devices.
7. **Special expertise:**
   Management of pesticide and pharmaceutical registration programs; extensive experience with FIFRA, TSCA, and FDA guidelines; complete nutritional labeling programs; development, modification, and validation of analytical methods; in vivo kinetic disposition and metabolic fate of radiolabeled and nonradiolabeled test chemicals.
8. **Size of lab:**
   300,000 square feet at principal facility; 1,000,000 square feet available at other facilities.
9. **Equipment available:**
   GC, GC/MS, AA, inductively coupled argon plasma atomic emission spectrophotometer, detectors and spectrophotometer amino acid analyzers.
10. **Key personnel:**
    Robert Conway, General Manager
    Robert Fischbeck, Director of Business Development
    Earl Richter, Director of Nutritional Chemistry
    Ihor Bekersky, Ph.D., Director of Research Chemistry
    Karen MacKenzie, Ph.D., Director of Toxicology
    Ronald Larson, Director of Toxicology Operations
    Byron Boysen, D.V.M., Director of Pathology
11. **Staff:**
    Professional and support at principal facility: 700, professional and support at other facilities: 2,000.
12. **GLP compliance:**
    Yes. Date of last inspection: August 2, 1988.
13. **FDA approval:**
14. **Accreditation:**
    AAALAC, U.S. EPA CLP Program.
15. **Communication capabilities:**
    Telephone number: (608) 241-4471
    Telefax number: (608) 241-7227
    Telex number: 703956 HAZRALMDS UD
    Computer availability: State of the art computerized tracking and reporting capabilities. Several analytical instruments are interfaced directly with a Hewlett Packard mainframe computer and data are stored using the Laboratory Automation System.
       Clients with the required hardware and software are able to interface directly with the computers for rapid exchange of data and information.
16. **Additional data:**
    Superfund site investigations, institutional review board, food packaging/interaction studies, regulatory affairs services. Additional facilities in Florida, Washington, DC, Maryland, Virginia, England, France, and Germany.
17. **Key contact:**
    Robert E. Conway

1. **Name of company:**
   Hazleton Laboratories America, Inc.
2. **Address:**
   900 Osceola Drive
   West Palm Beach, Florida 33409
3. **Corporate officers:**
   Donald P. Nielson
   Roy M. Dagnall
   Carl C. Schwan
   Robert E. Conway
4. **Services offered:**
   Safety testing, product efficacy and claims substantiation, dental studies, sunscreen testing, marketing research, and focus groups.
5. **Tests performed:**
   pH telemetry, oral irritation/sensitization studies in humans, fluoride effect on induced caries, denture cleanser and appliance effectiveness, repeat insult patch test in humans, phototoxicity/photoallergy studies in humans.
6. **Products tested:**
   Cosmetics, pharmaceuticals, antiperspirants, moisturizers, hair care products, personal hygiene products, malodor products, sunscreens, dental care products.
7. **Special expertise:**
   Periodontal disease, caries trials, pH telemetry, re-demineralization, pain control projects.
8. **Size of lab:**
   22,000 square feet at principal facility; 350,000 square feet at other facilities.
9. **Equipment available:**
   Solar simulators, environmental chamber, dental operatory, dental laboratory, beauty salon facilities, ophthalmoscope.
10. **Key personnel:**
    Charlene Bowman, Clinical Director
    Suru Mankodi, D.D.S., M.S.D., Director of Dental Sciences
    J. John Goodman, M.D., Medical Director
    Eileen D. Rutstein, Manager of Client Services
    Dyal C. Garg, Ph.D., Director of Clinical Research Unit
    Juan J. Martinez, Manager of Cosmetic and Proprietary Products
11. **Staff:**
    Professional and support at principal facility: 900, professional and support at other facilities: over 800.
12. **GLP compliance:**
    Institutional Review Board. Date of last inspection: December 27, 1988.
13. **FDA approval:**
14. **Accreditation:**
15. **Communication capabilities:**
    Telephone number: (407) 686-7210
    Telefax number: (407) 471-5295
16. **Additional data:**
    Regulatory affairs services, academic collaboration, independent quality assurance unit.
17. **Key contact:**
    Charlene Bowman

LISTING OF TOXICOLOGY LABORATORIES

1. **Name of company:**
   Hunter Environmental Services, Inc.
2. **Address:**
   Box 1703
   Gainesville, Florida 32602-1703
3. **Corporate officers:**
   Robert C. Minning
   Jack Doolittle
4. **Services offered:**
   Toxicity testing—fresh and saltwater organisms; eco-toxicology testing; complete analytical services.
5. **Tests performed:**
   Acute, early life stage, subchronics, chronics, phytotoxicity.
6. **Products tested:**
   Pesticides, industrial chemicals, drilling fluids, effluents, leachates, sediments, oils and disperants.
7. **Special expertise:**
8. **Size of lab:**
   Aquatic toxicology: 3,000 square feet; chemistry: 10,000 square feet.
9. **Equipment available:**
   GC, MS, scintillation counter, LC, ICAP, AA, IC.
10. **Key personnel:**
    G. Scott Ward
    Bruce A. Rabe
    C. Steve Manning
    Brian A. Wade
    Kim Rhodes
    Alex Brumfield
11. **Staff:**
    Toxicologists: 4, technicians: 4.
12. **GLP compliance:**
    Yes. Date of last inspection: March 1987, EPA Office of Compliance Monitoring; April 1988, EPA, Region IV.
13. **FDA approval:**
    N/A.
14. **Accreditation:**
15. **Communication capabilities:**
    Telephone number: (904) 332-3318
    Telefax number: (904) 332-0507
    Computer availability: Yes.
16. **Additional data:**
    Total Organic Halogens. Water and air monitoring and modeling, environmental engineering services.
17. **Key contact:**
    G. Scott Ward

1. **Name of company:**
   IIT Research Institute
2. **Address:**
   10 W. 35th Street
   Chicago, Illinois 60616-3799
3. **Corporate officers:**
   Dr. David Morrison, President
   Dr. Richard Ehrlich, Director of Life Sciences Research Department
4. **Services offered:**
   Toxicology testing and research, microbiology testing and research, immunology testing and research, analytical services, antiviral screening.
5. **Tests performed:**
   Acute, chronic, subchronic, gene-tox; immunotoxicity; inhalation toxicity; rats, mice, guinea pigs, hamsters, rabbits; in vivo and in vitro studies; carcinogenicity, initiation/promotion.
6. **Products tested:**
   Pesticides, lubricants, petrochemicals, pharmaceuticals, industrial chemicals, recombinant products.
7. **Special expertise:**
   Inhalation toxicology, immunotoxicology, developmental toxicology, pharmacokinetics, molecular biochemistry.
8. **Size of lab:**
   80,000 square feet.
9. **Equipment available:**
   GC, MS, radial scintillation counter, HPLC, IR, UV, AA, ICAP, FTIR, XRF, XRD, hematology analyzer, clinical pathology data computer, inhalation chambers, fluorescent microscope, horizontal electrophoresis apparatus for nucleic acid samples.
10. **Key personnel:**
11. **Staff:**
    Staff scientists and technicians: 80.
12. **GLP compliance:**
    Yes. Date of last inspection: NTP, May 1988; EPA, October 1986.
13. **FDA approval:**
    Yes. Date of last inspection: July 1982.
14. **Accreditation:**
    AAALAC, approved by the Japanese MAFF. NTP 5-year master agreement for inhalation toxicology.
15. **Communication capabilities:**
    Telephone number: (312) 567-4000
    Telefax number: (312) 567-4577
    Telex number: 282472 IITRI CGO
    Computer availability: VAX main frame, modem hookup for information services—Medline, Toxline.
16. **Additional data:**
    Total Organic Halogens. Extensive research capabilities in biopesticides, antivirals, anti-AIDS compounds, immune modulators, cancer chemoprevention, and chemotherapy.
17. **Key contact:**
    Dr. James D. Fenters

LISTING OF TOXICOLOGY LABORATORIES

1. **Name of company:**
   ImmuQuest Laboratories, Inc.
2. **Address:**
   13 Taft Court, Suite 200
   Rockville, Maryland 20850
3. **Corporate officers:**
   James L. McCoy, President
   Josie A. McCoy, CFO
4. **Services offered:**
   Immunotoxicology testing, skin sensitization-allergy testing in animals.
5. **Tests performed:**
   In vitro immunotoxicology, in vivo immunotoxicology in mice, rats, humans, guinea pigs.
6. **Products tested:**
   Immunomodulators (anti-AIDS therapeutic, immunosuppressive and immunostimulating agents), cosmetics, pesticides, personal care product ingredients (skin sensitization in guinea pigs).
7. **Special expertise:**
   Immunotoxicology, skin sensitization testing.
8. **Size of lab:**
   10,000 square feet.
9. **Equipment available:**
   Scintillation counter, ultracentrifuges, laminar flow hoods, ELISA reader, fluorescent microscopy.
10. **Key personnel:**
    James L. McCoy, Ph.D.
    H. J. Cher, M.D.
    Gregory Fiscbeth, B.S.
    Karyn Howard, B.S.
    Mary Wesoloski, A.A.
11. **Staff:**
    12.
12. **GLP compliance:**
    Yes. Date of last inspection: June 1987.
13. **FDA approval:**
14. **Accreditation:**
    USDA, OPRR.
15. **Communication capabilities:**
    Telephone number: (301) 762-7193
16. **Additional data:**
    Total Organic Halogens.
17. **Key contact:**
    James L. McCoy

1. **Name of company:**
   Kemron Environmental Services, Inc.
2. **Address:**
   755 New York Avenue
   Huntington, New York 11743
3. **Corporate officers:**
   Juan J. Gutierrez, President
   Mark Wilber, Vice President
4. **Services offered:**
   Industrial hygiene consulting and analytical services, chemical inventories/MSDS, environmental and workplace audits, hazardous communication training services.
5. **Tests performed:**
   Comprehensive sampling and analysis for asbestos, metals, organics, free silica, toxic gases, florides, EtO/N_2O, blood lead, BTX, and inorganics.
6. **Products tested:**
7. **Special expertise:**
   Light microscopy.
8. **Size of lab:**
   1,200 square feet.
9. **Equipment available:**
   GC/MS, GC, HPLC, AA, TOC, TOX, phase contrast microscope, polarized light microscope, x-ray diffractor.
10. **Key personnel:**
    Bernard P. Orlan, Regional Manager/Director of Technical Services
    Lisa G. Alianti, Manager of Field Services
    Jose Sosa, Manager of Engineering Services
    Jane Weber, Assistant Manager of Remediation Services
    Lisa DeVeau-Horton, Quality Control Coordinator
    Patricia LaMar-Kirkland, Laboratory Manager
    Maureen J. Nimphius, Laboratory Manager
11. **Staff:**
    Industrial hygienists: 5, microscopists: 20, industrial hygiene technicians: 10, chemists: 3, project managers: 5.
12. **GLP compliance:**
    New York State Department of Health, AIHA. Date of last inspection: October 1988, January 1988.
13. **FDA approval:**
14. **Accreditation:**
    AIHA, EPA, CDC, NYSDOH, NVLAP.
15. **Communication capabilities:**
    Telephone number: (516) 427-0950
    Telefax number: (516) 427-0989
16. **Additional data:**
    Total Organic Halogens.
17. **Key contact:**
    Maureen J. Nimphius

LISTING OF TOXICOLOGY LABORATORIES

1. **Name of company:**
   Laboratory Research Enterprises, Inc.
2. **Address:**
   6321 South 6th Street
   Kalamazoo, Michigan 49009-9611
3. **Corporate officers:**
   Robert W. Denison, President/CEO
   Martin R. Gilman, Ph.D., Vice President, Research
   Nicolas J. Schmelzer, Vice President, Sales and Marketing
4. **Services offered:**
   Toxicologic, reproductive, efficacy, pet food efficacy, testing specializing in dogs, cats, and rabbits.
5. **Tests performed:**
   Acute, subchronic, and chronic toxicities, reproduction studies in dogs.
6. **Products tested:**
   Studies conducted for the chemical, petrochemical, pharmaceutical, personal care, and pet food industries.
7. **Special expertise:**
   Male and female canine reproduction studies, including semen collection, counting, and evaluation; surgical procedures with accompanying recovery monitoring.
8. **Size of lab:**
   5,500 square feet limited-access research building plus an additional 100,000 square feet available for nonrestrictive testing.
9. **Equipment available:**
   Electrocardiography, surgical suite.
10. **Key personnel:**
    Martin R. Gilman, Ph.D., Toxicologist
    Sean D. Hughes, Ph.D., Veterinarian
    Anna Gall, D.V.M., Veterinarian
    Carol Johnson, B.S., Laboratory Supervisor
11. **Staff:**
    Toxicologists: 1, veterinarians: 2, technicians: 6.
12. **GLP compliance:**
13. **FDA approval:**
14. **Accreditation:**
    AAALAC accreditation 1987, member AALAS.
15. **Communication capabilities:**
    Telephone number: (616) 375-0482
    Telefax number: (616) 375-8403
16. **Additional data:**
    Total Organic Halogens.
17. **Key contact:**
    Martin R. Gilman, Ph.D.

1. **Name of company:**
   Lancaster Laboratories, Inc.
2. **Address:**
   2425 New Holland Pike
   Lancaster, Pennsylvania 17601
3. **Corporate officers:**
   Dr. Earl Hess, President/Founder
   Dr. Glenn Cahilly, Vice President/Director Support Services
   Dr. Fred Albright, Senior Vice President/Director Health Sciences Division
   Dr. J. Wilson Hershey, Vice President/Director Environmental Science Division
   Mr. Ken Hess, B.S., Vice President/Director of Finance and Data Processing
   Mrs. Carol Miller, M.B.A., Vice President/Director Human Resources
4. **Services offered:**
   Complete analytical services to the environmental, industrial hygiene, food, and pharmaceutical industry.
5. **Tests performed:**
   Food analyses, pharmaceutical analyses, industrial hygiene, environmental analyses, environmental profiles.
6. **Products tested:**
   Pesticides, cosmetics, pharmaceuticals, industrial chemicals, foods, asbestos, indoor air quality, health and safety, microbiology, water analysis, environmental.
7. **Special expertise:**
   Analytical chemistry, analytical method development, microbiology.
8. **Size of lab:**
   78,000 square feet.
9. **Equipment available:**
   GC/MS, GC, HPLC, AA, ACP, UV/VIS.
10. **Key personnel:**
    Sandra R. Bailey
    Nancy J. Bornholm
    Christine K. Brennian
    Richard M. Burke
    Eric W. Cuba
11. **Staff:**
    Chemists, microbiologists, technicians, support personnel: 300.
12. **GLP compliance:**
    Yes. Date of last inspection: October 21, 1986.
13. **FDA approval:**
    Site registration no. 2513291, October 21, 1986 and July 1988. Date of last inspection: DEA site registration no. PLO 213556 (schedules I–IV).
14. **Accreditation:**
    AALA, AIHA, AMA, AWWA, AOAC, BBB/Eastern PA, NFIB, ACIL, PMCA.
15. **Communication capabilities:**
    Telephone number: (717) 656-2301
    Telefax number: (717) 656-2681
    Computer availability: Lancaster Laboratories, Inc. Client Access System (CAS) allows almost any standard teleprinter or PC with modem to access our Sample Management System.
16. **Additional data:**
17. **Key contact:**
    Randall H. Guthrie

LISTING OF TOXICOLOGY LABORATORIES

1. **Name of company:**
   Langston Laboratories, Inc.
2. **Address:**
   2006 West 103rd Terrace
   Leawood, Kansas 66206
3. **Corporate officers:**
   C. Walter Langston, President
   Christina M. Scharff, Vice President
4. **Services offered:**
   Independent chemical and bacteriological testing.
5. **Tests performed:**
   Standard methods for environmental, microbiology, chemistry, food and agriproducts.
6. **Products tested:**
   All products.
7. **Special expertise:**
   Wastewater, hazardous waste, microbiology, analysis by standard methods.
8. **Size of lab:**
   16,000 square feet.
9. **Equipment available:**
   GC, GC/MS, IR, HPLC, GPC, UV-VIS, AA, ICP, Zeeman, TOC, TOX, fluorescence.
10. **Key personnel:**
    Don P. Miller
    Don Wright
    Gary Foushee
    Judith Russell
    David Beem
    Marsha Carroll
    Brian Smith
11. **Staff:**
    Analytical chemists: 6, CIH: 1, microbiologists: 2, technicians: 2.
12. **GLP compliance:**
13. **FDA approval:**
14. **Accreditation:**
    CLP-EPA: Kansas, Missouri, Nebraska; OSHA; NIOSH.
15. **Communication capabilities:**
    Telephone number: (913) 341-7800
16. **Additional data:**
    Analytical.
17. **Key contact:**
    Don P. Miller

1. **Name of company:**
   La Rocca Science Laboratories, Inc.
2. **Address:**
   1 Nell Court
   Dumont, New Jersey 07628
3. **Corporate officers:**
   Rudolph La Rocca, President
   Felicia La Rocca, Secretary/Treasurer
4. **Services offered:**
   Microbiology of foods, pharmaceuticals, beverages, medical devices, water, air; microbiological challenge studies; microbial evaluation of sanitizers, disinfectants, surgical scrubs (glove juice tests) process control; HACCP studies; GMP-training programs; process audits; sterility testing.
5. **Tests performed:**
   Microbiological—routine testing and research studies.
6. **Products tested:**
   Foods, beverages, disinfectants, sterility tests, pharmaceuticals, cosmetics, sanitizers, air, water, medical services, surgical scrub products.
7. **Special expertise:**
   Microbiological process analyses—problem-finding and solutions; hazard analyses—critical control points.
8. **Size of lab:**
   2,000 square feet.
9. **Equipment available:**
   Standard microbiological laboratory equipment, clean room, laminar flow hoods, steam autoclaves, millipore apparatus, air samplers, microscopes.
10. **Key personnel:**
    Mary Anne La Rocca, Ph.D.
    Paul La Rocca, Ph.D.
    Marion Corwin
11. **Staff:**
    Pharmacologist/Toxicologist: 1, microbiologists: 3, technicians: 5.
12. **GLP compliance:**
13. **FDA approval:**
    I.D. no. 2219396. Date of last inspection: January 20, 1988.
14. **Accreditation:**
    AALA, ACIL.
15. **Communication capabilities:**
    Telephone number: (201) 384-8509
16. **Additional data:**
    Analytical.
17. **Key contact:**
    Rudolph La Rocca

1. **Name of company:**
   Leberco Testing, Inc.
2. **Address:**
   123 Hawthorne Street
   Roselle Park, New Jersey 07204-0206
3. **Corporate officers:**
   Dr. Edwin C. Rothstein, President
   Brenda Rothstein, Secretary/Treasurer
4. **Services offered:**
   Toxicology, microbiology, environmental testing, analytical services.
5. **Tests performed:**
   Acute; subchronic; mice, rats, guinea pigs, rabbits, cats, dogs.
6. **Products tested:**
   Pharmaceuticals, medical devices, cosmetics, household and industrial chemicals, pesticides, herbicides, drinking water, wastewater.
7. **Special expertise:**
   Inhalation toxicology, GC/MS.
8. **Size of lab:**
   14,000 square feet.
9. **Equipment available:**
   UV/VIS, IR, HPLC, GC, GC/MS.
10. **Key personnel:**
    E. C. Rothstein, Ph.D., Director
    E. Levy, Ph.D., Associate Director
    F. Kennedy, M.S., Manager, Quality Assurance
    C. Reilly, B.S., Manager Toxicology
    L. Ostonal, B.S., Manager, Analytical Services
11. **Staff:**
    Toxicologists: 5, microbiologists: 7, technicians: 5, chemists: 5.
12. **GLP compliance:**
    Yes. Date of last inspection: July 28, 1988.
13. **FDA approval:**
    Yes. Date of last inspection: June 1986.
14. **Accreditation:**
15. **Communication capabilities:**
    Telephone number: (201) 245-1933
    Telefax number: (201) 245-6253
    Computer availability: IBM PC clone network (3-COM).
16. **Additional data:**
    Total Organic Halogens. Safety assessment testing for the food, drug, cosmetic, and chemical industries on a contractual basis.
17. **Key contact:**
    Dr. E. C. Rothstein

1. **Name of company:**
   Arthur D. Little, Chemical & Life Sciences Section
2. **Address:**
   25 Acorn Park   30 Memorial Drive
   Cambridge, Massachusetts 02140   Cambridge, Massachusetts 02140
3. **Corporate officers:**
   Charles La Mantia, President/CEO
   John Ketteringham, Senior Vice President
   Anthony Graffeo, Manager, Chemical and Life Sciences
4. **Services offered:**
   Toxicology (in vivo, in vitro), efficacy, metabolism, immunology, cell biology/microbiology, regulatory sciences, analytical chemistry, histopathology, phototoxicity, combustion toxicology, inhalation toxicology, immunotoxicology, protocol review and development.
5. **Tests performed:**
   Acute, subchronic, chronic, reproduction, inhalation, combustion, in vivo, in vitro, mutagenicity, gene toxicology, ADME, pharmacokinetics, phototoxicology, immunotoxicology, analytical services (GC, HPLC, GC/MS), all conventional species—mice to monkeys.
6. **Products tested:**
   Pharmaceutical, chemical, agrichemical, petrochemical, cosmetic, flavors, fragrances, consumer products.
7. **Special expertise:**
   General toxicology, in vitro toxicology, phototoxicology, ADME, cutaneous toxicology, inhalation, combustion, analytical chemistry, metabolic fate, environmental affairs.
8. **Size of lab:**
   50,000 square feet; animal facility: 25,000 square feet.
9. **Equipment available:**
   GC, MS, HPLC, EM.
10. **Key personnel:**
    A. Sivak, Toxicology   P. Farrow, Analytical Chemistry
    C. Berman, Toxicology   T. Graffeo, Analytical Chemistry
    B. Stuart, Inhalation   P. Palm, Regulatory Affairs
    K. Loveday, Cell Biology   D. Hayes, Quality Assurance
    M. Chadwick, Metabolism   J. Swiniarski, Animal Facilities
    A. Nomeir, Metabolism   J. Fox, Veterinarian
    L. Yelle, Analytical Chemistry
11. **Staff:**
    Total: 100.
12. **GLP compliance:**
    Yes. Date of last inspection: 1987.
13. **FDA approval:**
    Yes. Date of last inspection: 1987.
14. **Accreditation:**
    AAALAC, AALAS, MSPCA, USDA.
15. **Communication capabilities:**
    Telephone number: (617) 864-5770
    Telefax number: (617) 547-3617
    Telex number: 921436
16. **Additional data:**
    Total Organic Halogens.
17. **Key contact:**
    Jack Pulidoro, Ph.D.

LISTING OF TOXICOLOGY LABORATORIES

1. **Name of company:**
   Lycott Environmental Research, Inc.
2. **Address:**
   600 Charlton Street
   Southbridge, Massachusetts 01550
3. **Corporate officers:**
   Lee D. Lyman
4. **Services offered:**
   Analysis of liquids and solids for EPA priority pollutants, nutrients, minerals; NPDES compliance monitoring.
5. **Tests performed:**
   Chemical: GC/MS, GC, AA, calorimetric, gravimetric. Biological: bacteria, phytoplankton.
6. **Products tested:**
   Most liquids and solids.
7. **Special expertise:**
   EPA priority pollutants.
8. **Size of lab:**
   3,000 square feet.
9. **Equipment available:**
   GC/MS, GC-ECD, GC-FID, GC-Hall, AA, AA-furnace, AA-cold vapor, AA-hydride, spectrophotometer.
10. **Key personnel:**
    Ron LaBlanc, Lab Manager
    Jim Baril, GC/MS Chemist
    Linda Sylvester, AA Chemist
    Monique Beaupre, GC Chemist
11. **Staff:**
    Chemists: 4, technicians: 3.
12. **GLP compliance:**
13. **FDA approval:**
14. **Accreditation:**
    Certified in Massachusetts, Connecticut, Rhode Island.
15. **Communication capabilities:**
    Telephone number: (508) 765-0101
16. **Additional data:**
    Analytical. Expert hydrological services, site assessments, environmental impact studies, lake and pond management, professional engineering services.
17. **Key contact:**
    Ron LeBlanc

1. **Name of company:**
   Malcolm Pirnie, Inc.
2. **Address:**
   2 Corporate Park Drive   100 Grasslands Road
   White Plains, New York 10602 Elmsford, New York 10523
3. **Corporate officers:**
   Paul L. Busch, Ph.D., P.E./President
   Jeffrey M. Lauria, Ph.D., P.E./Vice President of Laboratory Services
   John C. Henningson, P.E./Vice President of Environmental Sciences
4. **Services offered:**
   Ecotoxicology: aquatic toxicity tests; analytical: water, wastewater, soils, sludge.
5. **Tests performed:**
   Phototoxicity (algae and aquatic plants); toxicity to invertebrates and fish, acute and short-term chronic; freshwater and marine species.
6. **Products tested:**
   Any products covered by TSCA and FIFRA; wastewater effluents.
7. **Special expertise:**
   Culture and testing of aquatic plants for FIFRA, subpart J, and TSCA guideline 797.1050.
8. **Size of lab:**
   3,000 square feet.
9. **Equipment available:**
   GC, GC/MS, AA, TOC, spectrophotometer, coulter counter.
10. **Key personnel:**
    Jane Hughes, Director, Aquatic Toxicology Program
    Meryl Alexander, Principal Scientist, Aquatic Toxicology
    Eugene Cocozza, Manager, Analytical Laboratory
11. **Staff:**
    Aquatic toxicology: 4, analytical chemistry: 5.
12. **GLP compliance:**
    Yes. Date of last inspection: April 21, 1987.
13. **FDA approval:**
    N/A.
14. **Accreditation:**
    Certified for water and wastewater analyses by New York Department of Health, New Jersey Department of Environmental Protection, and Connecticut Department of Health Services.
15. **Communication capabilities:**
    Telephone number: (914) 347-2970 (analytical), (914) 347-2974 (aquatic)
    Computer availability: Access to several major literature search and database services—DIALOG, CJS.
16. **Additional data:**
    Full services in environmental biology, chemistry, and engineering, field surveys; environmental audits; environmental hazard evaluation and risk assessment; environmental fate modeling; prediction of exposure concentrations.
17. **Key contact:**
    Jane S. Hughes

LISTING OF TOXICOLOGY LABORATORIES

1. **Name of company:**
   M. B. Research Laboratories, Inc.
2. **Address:**
   Steinberg and Wentz Roads
   Box 178
   Spinnerstown, Pennsylvania 18968
3. **Corporate officers:**
   Oscar M. Moreno, President
   Mary Teresa Moreno, Vice President
4. **Services offered:**
   Toxicology testing.
5. **Tests performed:**
   Acute, subchronic, chronic.
6. **Products tested:**
   All types of products from all industries.
7. **Special expertise:**
8. **Size of lab:**
   16,000 square feet.
9. **Equipment available:**
10. **Key personnel:**
    Daniel R. Cerven, M.S.
11. **Staff:**
    17.
12. **GLP compliance:**
    Yes. Date of last inspection: August 1988.
13. **FDA approval:**
    Yes. Date of last inspection: August 1988.
14. **Accreditation:**
    AAALAC.
15. **Communication capabilities:**
    Telephone number: (215) 536-4110
    Telefax number: (215) 536-1816
16. **Additional data:**
    Total Organic Halogens.
17. **Key contact:**
    Oscar M. Moreno, Ph.D.

1. **Name of company:**
   Microbiological Associates, Inc.
2. **Address:**
   9900 Blackwell Road
   Rockville, Maryland 20850
3. **Corporate officers:**
   Lew Schuster, President
4. **Services offered:**
   Health sciences testing and services company; safety testing on chemical and biological materials; development and validation of new test methods for determining the biological effects of potentially hazardous materials; diagnostic testing for animal diseases.
5. **Tests performed:**
   General toxicology: acute, subchronic, chronic, carcinogenicity, teratology/reproductive toxicology; genetic toxicology: bacterial and mammalian mutagenesis, in vitro and in vivo cytogenetics, cell transformation, DNA damage and repair; biotechnology: process validation, characterization of recombinant cell lines and hybridomas, final product tests; molecular virology: major contractor for HIV to NIH, commercial HIV services.
6. **Products tested:**
   Industrial chemical, agricultural chemical, petroleum products, pharmaceuticals, personal care products, consumer products, medical devices, biotechnology products.
7. **Special expertise:**
   Nose-only inhalation, skin-painting, biotechnology product safety testing.
8. **Size of lab:**
   85,000 square feet.
9. **Equipment available:**
10. **Key personnel:**
    Li L. Yang, Director of Marketing
    Rodger Curren, Director of Research
    James Fenno, Director of Biotechnology
    Tom Mulligan, Director of Toxicology
    David Jacobson-Kram, Director of Genetic Toxicology
11. **Staff:**
    190 people.
12. **GLP compliance:**
    Yes. Date of last inspection: Fall 1987.
13. **FDA approval:**
    Yes. Date of last inspection: Fall 1987.
14. **Accreditation:**
    AAALAC, JMAFF, AALAS, TCA.
15. **Communication capabilities:**
    Telephone number: (301) 738-1000
    Telefax number: (301) 738-1036
    Telex number: 908793
16. **Additional data:**
    Total Organic Halogens.
17. **Key contact:**
    Cheryl Respess

LISTING OF TOXICOLOGY LABORATORIES

1. **Name of company:**
   Midwest Research Institute
2. **Address:**
   425 Volker Boulevard
   Kansas City, Missouri 64110
3. **Corporate officers:**
   J. C. McKelvey, President/CEO
   H. M. Hubbard, Executive Vice President/Director, SERI
   J. A. Dinwiddle, Senior Vice President/CEO/Corporate Secretary
   M. C. Kirk, Treasurer/Executive Director, Finance
   D. N. Sunderman, Senior Vice President/Director of Kansas City Operations
   J. L. Spigarelli, Vice President/Associate Director of Kansas City Operations
4. **Services offered:**
   Toxicology research, pharmacological metabolism, disposition and pharmacokinetic studies, analytical (pharmaceutical) chemistry, microbiological analysis.
5. **Tests performed:**
   Acute, subacute, chronic toxicity in rodents, dogs, and primates; metabolism studies in rodents, dogs, and monkeys; genotoxicity assays with bacterial and mammalian cells; pharmaceutical chemical analysis including method development, characterization, drug impurity identification; parent and metavolite identification of chemicals in body fluids and tissues; bacterial and fungal toxins in foods and environmental samples.
6. **Products tested:**
   Bulk drugs and formulations, experimental drugs, industrial chemicals, foods, pesticides, environmental samples.
7. **Special expertise:**
   Pharmacokinetics, in vitro toxicological methods development, pharmaceutical chemistry assay development, drug fate and distribution studies, chemical analyses (HPLC, GC, GC/MS, LC/MS, NMR, TLC, IR, UV, VIS spectroscopy), bacterial ecology.
8. **Size of lab:**
   228,500 square feet; Deramus, Missouri, field station: 17,500 square feet.
9. **Equipment available:**
   HPLC, GC, MS, AA, NMR, FTIR, liquid gamma scintillation counters, microscopes, animal facilities.
10. **Key personnel:**

Monaem El-Hawari, Ph.D.	Rakesh Dixit, Ph.D.
Evelyn Murrill, Ph.D.	Gene Ray, B.S.
Frank Wells, Ph.D.	Gene Podrebarac, Ph.D.,
Maxine Stoltz, B.S.	Quality Assurance

11. **Staff:**
    Environmental chemists: 29, environmental scientists: 32, toxicologists/biologists: 16, chemists: 72, chemical engineers: 24, behavioral scientists: 9, bioanalytical chemists: 49, environmental engineers: 20, mechanical engineers: 8.
12. **GLP compliance:**
    Yes. Date of last inspection: November 1985.
13. **FDA approval:**
    Yes. Date of last inspection: November 1985.
14. **Accreditation:**
    AAALAC.
15. **Communication capabilities:**
    Telephone number: (816) 753-7600
    Telefax number: (816) 753-8420
    Telex number: 910-771-2128
    Computer availability: Agreements are in effect with numerous information vendors and agencies for on-line access to databases. Specialized computer networks are developed for direct transmission of data and reports to clients.
16. **Additional data:**
    Total Organic Halogens.
17. **Key contact:**
    Margaret Gunde

1. **Name of company:**
   Nebraska Testing Corporation
2. **Address:**
   4453 South 67th Street
   Omaha, Nebraska 68117
3. **Corporate officers:**
   Daniel E. McCarthy, Vice President/General Manager
   George C. Phelps, Manager, Chemical and Industrial Services
4. **Services offered:**
   Civil engineering: material and construction testing; Industrial: manufacturing product testing; nondestructive testing; Scientific: chemical, microbiology, environmental, industrial hygiene.
5. **Tests performed:**
6. **Products tested:**
7. **Special expertise:**
   Nondestructive testing, environmental assessments, subsurface drilling/exploration, soil mechanics testing, microbiological evaluations, industrial hygiene, asbestos abatement surveys/specifications.
8. **Size of lab:**
   22,000 square feet.
9. **Equipment available:**
   GC, UV spectrophotometer, AA spectrophotometer, subsurface interface radar (ground penetrating), ultrasonic test equipment, x-ray and gamma-ray radiography, radiation survey meters, truck-mounted drill rigs, universal tension/compression machine.
10. **Key personnel:**
    Al Rahman
    Donald F. Stevens
    Lynn A. Knudtson
    Brian A. Stemmermann
    Bill Willis
    John Selmer
11. **Staff:**
    Civil engineers: 4, microbiologists: 2, geologists: 3, technicians: 20, chemists/chemical engineers: 4.
12. **GLP compliance:**
    N/A.
13. **FDA approval:**
    Registration no. 1923996. Date of last inspection: July 29, 1986. Drug Enforcement Agency no. PN0244208.
14. **Accreditation:**
    USDA, LAB no. 3185: meat testing; EPA LAB no. 7420: (Asbestos); NIOSH PAT no. 68106-001; state of Nebraska water testing.
15. **Communication capabilities:**
    Telephone number: (402) 331-4453
    Telefax number: (402) 331-5961
16. **Additional data:**
    Analytical.
17. **Key contact:**
    George C. Phelps

LISTING OF TOXICOLOGY LABORATORIES

1. **Name of company:**
   North American Science Associates, Inc.
2. **Address:**
   2261 Tracy Street
   Northwood, Ohio 43619

   9 Morgan
   Irvine, California 92718
3. **Corporate officers:**
   Theodore W. Gorski, Ph.D., Chairman
   Richard F. Wallin, D.V.M., Ph.D., President
   Lucille F. Gorski, Executive Vice President
4. **Services offered:**
   Toxicology, microbiology, chemistry, clinical research.
5. **Tests performed:**
   Acute toxicity (per USP, FHSA, other regulatory guidelines) in vivo with rabbits, rodents, primates, and canines; in vitro tests–agarose overlay, MEM elution, inhibition of cell growth, Ames test; sterility, bioburden, other microbial tests; IR/FTIR, gas and liquid chromatography; USP assays; pharmaceutical testing (packaging and dissolution testing).
6. **Products tested:**
   Medical devices and device components, raw materials, cosmetics, specialty chemicals, pharmaceuticals.
7. **Special expertise:**
   Short-term acute toxicology, intraocular implants and other surgical procedures, in vitro tests.
8. **Size of lab:**
   96,000 square feet.
9. **Equipment available:**
   GC/HPLC, IR/FTIR, UV-VIS, electron microscope.
10. **Key personnel:**
    Richard F. Wallin, D.V.M., Ph.D., President/Scientific Director
    William C. Bradbury, Ph.D., General Manager, California
    Paul J. Upman, B.A., Manager, Toxicology, Ohio
    Paul Rudko, M.S., Manager, Toxicology, California
    Laurel E. Stroempl, B.S., Corporate Manager, Quality Assurance
    R. Douglas Hume, Ph.D. Manager, Microbiology Chemistry, Clinical Research, Ohio
    John J. Broad, B.S., M.B.A., Manager, Microbiology, California
    Carolyn Schultz, M.S., Manager, Chemistry, California
11. **Staff:**
    Study directors, managers, supervisory personnel: 34, technicians/technologists: 66, other: 35.
12. **GLP compliance:**
    Yes. Date of last inspection: December 1985.
13. **FDA approval:**
    N/A.
14. **Accreditation:**
    AAALAC, AALAS, CTFA.
15. **Communication capabilities:**
    Telephone number: (419) 666-9455
    Telefax number: (419) 666-2954
    Telex number: 810-442-1732

    (714) 951-3110
    (714) 951-3280
    628-29106
16. **Additional data:**
    Total Organic Halogens. Manufacture HBAC thromboresistant coating, distribute chemical sterilization indicators for ETO and radiation, manufacture sportrol® biological indicators.
17. **Key contact:**
    Suellen R. Romick

1. **Name of company:**
   Northview Pacific Laboratories, Inc.
2. **Address:**
   2800 Seventh Street
   Berkeley, California 94710

   1880 Holste Road
   Northbrook, Illinois 60062
3. **Corporate officers:**
   Martin Spalding, President
   Edward Kelly, Vice President
4. **Services offered:**
   Toxicology, microbiology, and sterility assurance testing; analytical laboratory services; tissue culture; virology; immunology.
5. **Tests performed:**
   Acute toxicology, in vivo (guinea pigs, rabbits, mice, rats), in vitro, biocompatiblity studies, EPA and FDA registrations.
6. **Products tested:**
   Medical device/hospital supply, drugs and pharmaceuticals, biotechnology, ophthalmic goods, specialty chemicals, food and vitamins, cosmetics.
7. **Special expertise:**
   EPA registration, stability studies, studies on contact lenses and solutions.
8. **Size of lab:**
   California facility: 10,000 square feet; Illinois facility: 15,000 square feet.
9. **Equipment available:**
   GC, HPLC, AA, IR, UV/VIS, permeometer.
10. **Key personnel:**
    Mary Jane Deenihan, Manager, Toxicology/Immunology Department
    Paul Adlaf, Ph.D., Manager, Analytical Services
    Odin Ansari, Manager, Microbiology Department
    Clyde Goodheart, M.D., Ph.D., Manager, Tissue Culture/Virology
    Tom Spalding, Operations Manager
11. **Staff:**
    Toxicologists: 2, microbiology technicians: 8, toxicology technicians: 8, chemists: 10, microbiologists: 5.
12. **GLP compliance:**
    Yes. Date of last inspection: July 1987.
13. **FDA approval:**
    Yes. Date of last inspection: June 1987.
14. **Accreditation:**
    Self-accredited animal facility, assurance on file with NIH; several staff members accredited by AALAS.
15. **Communication capabilities:**
    Telephone number: (415) 548-8440
    Telefax number: (415) 548-5425
16. **Additional data:**
    Total Organic Halogens.
17. **Key contact:**
    Joanne Spalding

1. **Name of company:**
   Oak Creek Laboratory of Biology, Department of Fisheries & Wildlife
2. **Address:**
   Nash Hall, Room no. 104
   Oregon State University
   Corvallis, Oregon 97331-3803
3. **Corporate officers:**
   Lawrence R. Curtis, Ph.D.
   William J. Liss, Ph.D.
   Wayne K. Seim, M.S.
   Hilliary M. Carpenter
4. **Services offered:**
   Aquatic toxicology and ecotoxicology with analytical support.
5. **Tests performed:**
   Acute 76-hour LC50 and growth studies with rainbow trout, bluegill sunfish, or fathead minnow; acute lethality and reproductive tests with Daphnia; embryolarval tests with rainbow trout (90 days).
6. **Products tested:**
   Pesticides and industrial chemicals.
7. **Special expertise:**
   Pharmacokinetics; xenobiotic metabolism by aquatic organisms.
8. **Size of lab:**
   9,764 square feet.
9. **Equipment available:**
   GC, HPLC, TLC, liquid scintillation counter, UV, visible spectrophotometer.
10. **Key personnel:**
    Beth Frederickson
    Bob Hoffman
    Chris Frissell
11. **Staff:**
    Toxicologists: 2, technicians: 3, ecologists: 1, graduate students: 6, water pollution biologists: 1.
12. **GLP compliance:**
    Specific compliances for particular USEPA and NIEHS grants and cooperative agreements.
13. **FDA approval:**
    None.
14. **Accreditation:**
    Member of the Society of Toxicology.
15. **Communication capabilities:**
    Telephone number: (503) 754-4531
    Telefax number: (503) 754-2400
    Telex number: 510-596-0682 OSU COVS
16. **Additional data:**
    Total Organic Halogens.
17. **Key contact:**
    Lawrence R. Curtis

1. **Name of company:**
   Product Safety Laboratories
2. **Address:**
   725 Cranbury Road
   East Brunswick, New Jersey 08816
3. **Corporate officers:**
   Ralph Shapiro, President
4. **Services offered:**
   Toxicology testing, analytical services, nutritional testing, clinical testing, genetic toxicology.
5. **Tests performed:**
   Acute and subchronic animal toxicology (rabbits, rodents), nutritional and feeding studies, bioassays, caloric equivalents, analytical testing, clinical studies (skin products, sunscreens).
6. **Products tested:**
   Cosmetics, industrial chemicals, pesticides, household products, pharmaceuticals.
7. **Special expertise:**
   Acute inhalation toxicology (EPA, OECD), cosmetic and personal care products, nutritional and feeding studies, bioassays, analysis of foods, phototoxicology, photoallergenicity.
8. **Size of lab:**
   14,000 square feet.
9. **Equipment available:**
   GC, HPLC, Andersson cascade impactor, IR vapor analyzer, UV irradiation equipment.
10. **Key personnel:**
    Ralph Shapiro, Ph.D., Toxicology
    Catherine Wo, Ph.D., Analytical
11. **Staff:**
    25.
12. **GLP compliance:**
    EPA. Date of last inspection: 1988.
13. **FDA approval:**
    Satisfactory. Date of last inspection: 1988.
14. **Accreditation:**
15. **Communication capabilities:**
    Telephone number: (201) 254-9200
    Telefax number: (201) 254-6736
    Computer availability: Word processing.
16. **Additional data:**
    A second division, Nutritional International, provides a full range of analytical, microbiological, and toxicological testing for the food industry.
17. **Key contact:**
    Walter Newman

LISTING OF TOXICOLOGY LABORATORIES

1. **Name of company:**
   Recon Systems, Inc.
2. **Address:**
   Route 202 North
   Box 460
   Three Bridges, New Jersey 08887
3. **Corporate officers:**
   Norman J. Weinstein, President
   Richard F. Toro, Executive Vice President
   Robert M. Wolfertz, Vice President
   William Bertele, Vice President
4. **Services offered:**
   Field sampling and analysis for environmental pollution parameters in air, gases, waters, soils.
5. **Tests performed:**
   Analytical services; RCRA hazardous waste determinations; water treatability studies for biochemical oxidation, air stripping, carbon adsorption, heavy metal precipitation; ultimate and proximate analysis; ASTM distillations.
6. **Products tested:**
   Mostly environmental samples.
7. **Special expertise:**
   GC, AA, IR.
8. **Size of lab:**
   7,000 square feet.
9. **Equipment available:**
   GC, AA, IR.
10. **Key personnel:**
    Frank W. Swetits, B.S., Air/Stack Testing
    G. Stephen Hornberger, B.S., Chemical Laboratory
    Joanne J. Wattstein, B.S., Senior Chemical Engineer
    C. David Ruff, B.S., Senior Chemical Engineer
    J. Douglas Reid-Green, M.S., Geologist
    Bernhard Meyer, B.A., Geologist
11. **Staff:**
    Managers: 7, technicians: 3, senior professionals: 7, associates: 8, staff professionals: 17.
12. **GLP compliance:**
    N/A.
13. **FDA approval:**
    N/A.
14. **Accreditation:**
    Certified by New Jersey Department of Environmental Protection for water and wastewater analysis.
15. **Communication capabilities:**
    Telephone number: (201) 782-5900
    Telefax number: (201) 788-0830
    Computer availability: IBM, IBM compatible PC, XT, AT microcomputers; access to main frame computer equipment with Princeton University computer center.
16. **Additional data:**
    Total Organic Halogens.
17. **Key contact:**
    Richard F. Toro

1. **Name of company:**
   Research Triangle Institute
2. **Address:**
   3040 Cornwallis Road
   Box 12194
   Research Triangle Park, North Carolina 27709
3. **Corporate officers:**
   George R. Herbert, President
   Dr. Daniel G. Horvitz, Executive Vice President
   R. S. McLean, Treasurer
   Suzanne Nash, Secretary
   Grace C. Boddie, Vice President/Senior Counsel
4. **Services offered:**
   Toxicology testing.
5. **Tests performed:**
   In-vivo toxicology/oncogenicity, cellular and genetic toxicology, mammalian genetics/mutagenesis, teratology/neurobehavioral toxicology, reproductive endocrinology/toxicology.
6. **Products tested:**
   Industrial chemicals, pharmaceuticals, cosmetics, environmental chemicals (pollutants), therapeutic agents, drugs of abuse.
7. **Special expertise:**
   Mutagenesis, carcinogenesis, teratology, reproductive toxicology, behavioral toxicology, general toxicology, genetic toxicology, endocrinology, pharmacokinetics, xenobiotic metabolism, metabolism after dermal exposure, electron microscopy, radiochemical synthesis, analytical method development (RIA, EIA, HPLC, GC, MS).
8. **Size of lab:**
   118,776 square feet.
9. **Equipment available:**
   GC, HPLC, CCC, EI, CI, FAB, GC/MS, HPLC/MS, CCC/MS, scintillation counters, sample oxidizer (for radiolabeling), nuclear magnetic resonance spectrophotometers, scanning and transmission EM.
10. **Key personnel:**
11. **Staff:**
    Toxicologists, Ph.D.: 9, technicians: 17, chemists, Ph.D.: 56, toxicologists, M.S.: 6, animal caretakers: 8, M.S.: 28; D.V.M.: 1, B.S.: 72, secretarial: 2.
12. **GLP compliance:**
    Research Triangle Institute has had a quality assurance/GLP compliance program since the implementation of the FDA/GLP in 1979.
13. **FDA approval:**
    Yes. Date of last inspection: March 17–21, 1986.
14. **Accreditation:**
    AAALAC since 1982.
15. **Communication capabilities:**
    Telephone number: (919) 541-6000
    Telefax number: (919) 541-6499
    Telex number: 802509 (RTI RTPK)
    Computer availability: Wide range of computer facilities available.
16. **Additional data:**
    Total Organic Halogens.
17. **Key contact:**
    Lisa C. Jeffrey

LISTING OF TOXICOLOGY LABORATORIES 57

1. **Name of company:**
   Ricerca, Inc.
2. **Address:**
   7528 Auburn Road
   Painesville, Ohio 44077
3. **Corporate officers:**
   James A. Scozzie, Ph.D., President
   Robert A. Baxter, Vice President
   Timothy E. Tinkler, J.D., Secretary
   Michael O. Dougherty, Treasurer
4. **Services offered:**
   Toxicology, animal metabolism, environmental fate, plant metabolism, residue analysis, analytical services, synthesis chemistry, agricultural research.
5. **Tests performed:**
   Acute oral toxicity, acute dermal toxicity, primary dermal irritation, primary eye irritation, dermal sensitization, subchronic toxicity, multigeneration/reproduction, chronic toxicity, tumorgenicity.
6. **Products tested:**
   Pesticides, herbicides, cosmetics, pharmaceuticals, biocontrol agents, biocides, plant growth regulants, industrial chemicals.
7. **Special expertise:**
   Assists chemical producers in basic research and development, synthesis chemistry, process development, test biological activity, full analytical support, residue analysis, environmental fate, toxicology, metabolism, registration assistance.
8. **Size of lab:**
   250,000 square feet.
9. **Equipment available:**
   CAP-ECD/FID/NPD/MS, GC-LC/MS, GPC/Prep, Ultratrace determiners of dioxins, PCB's, ICP, IC, AA, XRF/XRD, SEM-EDS-WDS/TEM/Optical, FTIR/NMR/UV-VIS, DSC/TGA/TMA; BET, scintillation counters, classical wet chemistry.
10. **Key personnel:**
    Richard L. Cryberg, Ph.D., Manager, Analytical Services
    Carol A. Duane, Manager, Information Services
    James C. Killeen, Ph.D., Associate Director, Toxicology and Animal Metabolism
    Alfred F. Marks, Manager, Environmental Sciences
    Larry J. Powers, Ph.D., Manager, Agricultural/Chemical/Fermentation, Research and Development
    John A. Resetar, Manager, Process Development
11. **Staff:**
    Scientists, engineers, and support personnel: 170.
12. **GLP compliance:**
    Exercised upon client request. EPA, date of last inspection: March 1985.
13. **FDA approval:**
    Yes. Date of last inspection: March 1987.
14. **Accreditation:**
15. **Communication capabilities:**
    Telephone number: (216) 357-3300
    Telefax number: (216) 354-4415
    Telex number: 196212RICERCA UT
16. **Additional data:**
    Total Organic Halogens. A broad-based technology company with demonstrated capabilities in the synthesis, analysis, pilot scale development, formulation, product registration, and commercial support of a wide range of chemicals and/or chemical-gased processes.
17. **Key contact:**
    Joseph A. Kohonoski

1. **Name of company:**
   Savannah Laboratories & Environmental Services, Inc.
2. **Address:**

Savannah Division	Mobile Division	Tallahassee Division
5102 LaRoche Avenue	3707 Cottage Hill Road	2820 Industrial Plaza Boulevard
Savannah, Georgia 31404	Mobile, Alabama 36609	Tallahassee, Florida 32301

3. **Corporate officers:**
   James W. Andrews, President
   Janette Davis Long, Vice President
   Barbara S. Andrews, Treasurer/Secretary
4. **Services offered:**
   Full-service environmental and analytical testing laboratory serving governmental and private sector clients.
5. **Tests performed:**
   Solid waste and environmental impact analysis, analysis of water, groundwater, wastewater, contaminated soil/water, leachates, and aquatic toxicology.
6. **Products tested:**
   Environmental samples.
7. **Special expertise:**
   Specializing in hazardous waste, dioxin, CLP protocol, groundwater, Appendix 9, tributyl tin, priority pollutants, bioassay and permit studies.
8. **Size of lab:**
   Georgia facility: 15,000 square feet; Alabama facility: 5,000 square feet; Florida facility: 8,000 square feet.
9. **Equipment available:**
   GC/MS, GC, HPLC, IC, AA, ICP, IR, TOC, TOX, UV/VIS.
10. **Key personnel:**

James W. Andrews, Ph.D.	Ernie Walton	Jim Sciple
Jesse L. Smith	Lisa McLeod	Michele Lersch
W. R. Robbins	H. Windom, Ph.D.	
Steve White	Thomas L. Stephens	
	Janet Pruitt	

11. **Staff:**
    Chemical technicians: 30, chemists (Ph.D.): 2, chemists (B.S./M.S.): 52, biologists: 10.
12. **GLP compliance:**
    N/A.
13. **FDA approval:**
    N/A.
14. **Accreditation:**
    Certified in Georgia, Alabama, South Carolina, Florida, Virginia, New York, Tennessee.
15. **Communication capabilities:**

Telephone number: (912) 354-7858	(205) 666-6633	(904) 878-3994
Telefax number: (912) 352-0165	(205) 666-6696	(904) 878-9504

    Computer availability: In process of adding an on-line computer service.
16. **Additional data:**
    Total Organic Halogens.
17. **Key contact:**
    Janette Davis Long

1. **Name of company:**
   Scientific Associates, Inc.
2. **Address:**
   6200 South Lindbergh Boulevard
   St. Louis, Missouri 63123
3. **Corporate officers:**
   J. Desmon Davies, Ph.D., President
   Robert H. Moulton, Senior Vice President
   Kenneth E. Giebe, Vice President
4. **Services offered:**
   Toxicology, chemical analysis, microbiological evaluation.
5. **Tests performed:**
   Toxicology: skin and eye irritation-pyrogen test, acute, subacute, subchronic, neuro, organisolates in dogs, cats, rabbits, guinea pigs, rats, mice. Large farm animal work on special request. Chemistry: all types of analytical chemistry. Microbiology: sterility, antibacterial effectiveness, antibiotics, bacterial load and contamination, identity of species.
6. **Products tested:**
   Pharmaceuticals, industrial chemicals, feeds, others.
7. **Special expertise:**
   Analytical chemistry, acute toxicology.
8. **Size of lab:**
   30,000 square feet.
9. **Equipment available:**
   HPLC, GC, AA, dissolution equipment, class 100 room for sterility.
10. **Key personnel:**
    Kenneth Giebe, Chemistry
    Larry Snopek, Chemistry
    Paul Steingruby, Toxicology
    Gary Barron, Toxicology
    Anne Pickels, Microbiology
    David Haywood, Microbiology
11. **Staff:**
    Toxicologists: 5, chemical technicians: 5, toxicological technicians: 4, microbiologists: 9, chemists: 27, microbiological technicians: 2.
12. **GLP compliance:**
    Yes. Date of last inspection: February 17, 1988.
13. **FDA approval:**
    Yes. Date of last inspection: November 10, 1986.
14. **Accreditation:**
    AALAS.
15. **Communication capabilities:**
    Telephone number: (314) 487-6776
    Telefax number: (314) 487-3991
16. **Additional data:**
    Total Organic Halogens.
17. **Key contact:**
    Robert H. Moulton

1. **Name of company:**
   Scientific Associates, Inc.
2. **Address:**
   6200 South Lindbergh Boulevard
   St. Louis, Missouri 63123
3. **Corporate officers:**
   Robert H. Moulton, Senior Vice President
   Kenneth Giebe, Vice President
   Arthur Scharle, Secretary/Treasurer
   Joan F. Styles, Assistant Secretary/Treasurer
4. **Services offered:**
   Short-term toxicology, USP biological analysis, chemical analysis, microbiological testing and analysis.
5. **Tests performed:**
   Subchronic toxicity, acute toxicity, pressure response, depresser response, USP bioanalysis, chemical analysis of all types, microbiological analysis—antibiotics, etc., preservative effectiveness, sterility, pyrogen testing.
6. **Products tested:**
   Pharmaceuticals, antimicrobials, pesticides, soaps, industrial chemicals, feeds—antibiotics and additives, cosmetics.
7. **Special expertise:**
   Chemical analysis and development of chemical analytical tests.
8. **Size of lab:**
   25,000 square feet.
9. **Equipment available:**
   GC, HPLC, dissolutiion apparatus.
10. **Key personnel:**
    Kenneth Giebe, Chemistry
    Larry Snopek, Chemistry
    Anne Pickles, Microbiology
    Lisa Suggs, Microbiology
    Paul Stenorgruby, Biological testing
    Gary Barron, Biological testing
11. **Staff:**
    Toxicologists: 5, microbiologists: 10, chemists: 25.
12. **GLP compliance:**
    Yes. Date of last inspection: EPA and FDA, October 10–16, 1987.
13. **FDA approval:**
    Yes. Date of last inspection: October 10–16, 1987.
14. **Accreditation:**
15. **Communication capabilities:**
    Telephone number: (314) 487-6776
    Telefax number: (314) 487-3991
    Telex Number: 447664
16. **Additional data:**
    Total Organic Halogens.
17. **Key contact:**
    Robert H. Moulton

LISTING OF TOXICOLOGY LABORATORIES

1. **Name of company:**
   Serco Laboratories
2. **Address:**
   1931 Westcounty Road C-2
   St. Paul, Minnesota 55113
3. **Corporate officers:**
   Carol A. Kuehn, President
4. **Services offered:**
   Analytical testing on water, wastewater, groundwater, drinking water, hazardous waste, PCBs, priority pollutants, pesticides, herbicides, hazardous substances list compounds, target compunds list.
5. **Tests performed:**
   Chemical analysis including PCBs, priority pollutants, target compound list, metals analysis, bacteriological tests.
6. **Products tested:**
   Waters, waste and wastewater, hazardous waste, process solutions and byproducts, oils, solvents, unknowns, air samples, soils, solids.
7. **Special expertise:**
   EPA methodologies for analysis of water, wastewater, hazardous wastes, soils, solids.
8. **Size of lab:**
   13,000 Square feet.
9. **Equipment available:**
   GC/MS, GC/EC, GC/FID, GC/HALL-PID, HPLC, AA furnace and flame.
10. **Key personnel:**
11. **Staff:**
    Analytical chemists: 5, environmental technicians: 3, biologists: 2, sampling experts: 2, environmental consultant: 1.
12. **GLP compliance:**
    None.
13. **FDA approval:**
    Yes. Date of last inspection: Spring 1988.
14. **Accreditation:**
    EPA water pollution and water supply evaluation studies, Minnesota certification for bacteriological testing, WI certification, ACIL, Twin Cities Round Robin (founders), ACS, American Society of Civil Engineers, AWWA, Minnesota Society of Professional Engineers, WPCF, WI Laboratory Association.
15. **Communication capabilities:**
    Telephone number: (612) 636-7173
    Telefax number: (612) 636-7178
    Computer availability: In-house.
16. **Additional data:**
    Analytical. Field sampling; field sampling equipmental rental; Bailer decontamination for groundwater sampling; environmental consulting regarding EPA regulations, NPDES and MWCC programs, process control.
17. **Key contact:**
    Carol A. Kuehn

1. **Name of company:**
   Shrader Analytical & Consulting Laboratories, Inc.
2. **Address:**
   3814 Vinewood
   Detroit, Michigan 48208
3. **Corporate officers:**
   Stephen Shrader, President
   Grover Shrader
   Marianne L. Shrader, Controller
4. **Services offered:**
   Analytical laboratory service—in particular, organic chemical identification or analysis.
5. **Tests performed:**
   GC, GC/MS, MS—both high and low resolution.
6. **Products tested:**
   Anything that is considered organic.
7. **Special expertise:**
   Mass spectrometry and specialized mass spectrometry software.
8. **Size of lab:**
   6,000 square feet.
9. **Equipment available:**
   GC, MS-30 GC/MS, VG 7070 GC/MS, Finnigan 3300 GC/MS.
10. **Key personnel:**
11. **Staff:**
    Organic Ph.D. chemists: 3, associates: 1, organic M.S.: 1, B.S.: 1.
12. **GLP compliance:**
13. **FDA approval:**
14. **Accreditation:**
15. **Communication capabilities:**
    Telephone number: (313) 894-4440
    Telefax number: (313) 894-4489
16. **Additional data:**
    Analytical.
17. **Key contact:**
    Marianne L. Shrader

1. **Name of company:**
   Skinner & Sherman Labs, Inc.
2. **Address:**
   300 Second Avenue
   Box 521
   Waltham, Massachusetts 02254
3. **Corporate officers:**
   Dr. John P. Appleton, President
   Dr. Haldean Dalzell, Vice President/Lab Director
4. **Services offered:**
   Environmental sample testing services; product testing and failure analysis; arson, accident, and other forensic investigations.
5. **Tests performed:**
   Chemical analysis and characterization—GC, GC/MS, FTIR, NMR, HPLC; optical microscopy; atomic absorption and emission; wet chemistry.
6. **Products tested:**
   water, soil, sludge, vegetation, animal tissue.
7. **Special expertise:**
   Trace inorganic and organic analysis.
8. **Size of lab:**
   8,000 square feet.
9. **Equipment available:**
   GC/MS, HPLC, AA, ICP, NMR, optical microscopy, weatherometers-UV, SO2, salt fog, ozone, thermal analyzers (TGA, DSC, TMA, DTA).
10. **Key personnel:**
11. **Staff:**
    Chemists: 20.
12. **GLP compliance:**
13. **FDA approval:**
    Yes. Date of last inspection: February 1988.
14. **Accreditation:**
    American Council of Independent Laboratories, Inc.; American Society for Testing & Materials; U.S. Environmental Protection Agency/CLP Program; various state certifications for testing services: Massachusetts, New York, Vermont, New Hampshire.
15. **Communication capabilities:**
    Telephone number: (617) 890-7200
    Telefax number: (617) 890-3883
16. **Additional data:**
    Analytical
17. **Key contact:**
    Dr. John P. Appleton

1. **Name of company:**
   Southern Research Institute
2. **Address:**
   2000 Ninth Avenue, South
   Box 55305
   Birmingham, Alabama 35255-5305
3. **Corporate officers:**
   John W. Rouse, Jr., President
   John A. Montgomery, Senior Vice President
   Paul Sharbel, Vice President/Treasurer
4. **Services offered:**
   Toxicology and carcinogenesis, anticancer drug development, genetics of multiple drug resistance, antiviral drug development, molecular genetics, cell biology, bacteriology and mycology; molecular virology, biochemistry research and development, analytical chemistry, controled-release technology, biotechnology, biochemical pharmacology, microbial and mammalian mutagenesis.
5. **Tests performed:**
   Rodent cytogenetic assays—in vitro, in vivo; rodent unscheduled DNA synthesis assays—in vitro, in vivo; rodent germ cell assay; in vitro assays using human cells; special cytogenetic services; engineering analysis.
6. **Products tested:**
   Hazardous and nonhazardous solid waste disposal, treatment, and reduction; wastewater treatment; air pollution control; incinerator effluents; analysis of dioxins, furans, and PCBs; trace inorganic and organic compounds.
7. **Special expertise:**
   Analytical and engineering capabilities; bioassays.
8. **Size of lab:**
   390,000 square feet.
9. **Equipment available:**
   AA, IC, SEM, SEM/ED x-ray, x-ray diffraction, GC/MS, LC/MS, high-resolution MS, fast-atom bombardment/MS/HPLC, TGA, DSC.
10. **Key personnel:**
    Dr. Dan Griswold, Director, Chemotherapy and Toxicology Research Department
    Dr. Donald Hill, Director, Biochemistry Research Department
    Dr. Herbert Miller, Director, Analytical and Physical Chemistry Research Department
    Dr. Grady Nichols, Director, Environmental Sciences Research Department
    Colteus Pears, Director, Mechanical and Materials Engineering Research Department
    Dr. John Secrist, Director, Organic Chemistry Research Department
11. **Staff:**
    700.
12. **GLP compliance:**
13. **FDA approval:**
14. **Accreditation:**
15. **Communication capabilities:**
    Telephone number: (205) 581-2000
16. **Additional data:**
    Total Organic Halogens
17. **Key contact:**
    J. Craig Battles

1. **Name of company:**
   South Mountain Laboratories, Inc.
2. **Address:**
   380 Lackawanna Place
   South Orange, New Jersey 07079
3. **Corporate officers:**
   C. N. Mangieri, President/Director
   M. G. Mangieri, Vice President/Secretary
4. **Services offered:**
   Small animal toxicology, pyrogen testing rabbit and LAL hormones; irritation—Draize skin irritation, eye and skin, cosmetics, HPLC, safety, implant-plastic; most biologics, anticarcinogenic, heparin, human skin, insulin, sterility.
5. **Tests performed:**
6. **Products tested:**
   Pharmaceuticals, cosmetics, anticarcinogenic pyrogens, hormonal, anticoagulants, insulin.
7. **Special expertise:**
   Pyrogens, hormones, developmental toxicology.
8. **Size of lab:**
   8,000 square feet.
9. **Equipment available:**
   HPLC, all bio-lab equipment, manometers, fever measuring equipment, tissue bath, incubators.
10. **Key personnel:**
11. **Staff:**
    Biologists: 20, statistician: 1, director: 1.
12. **GLP compliance:**
13. **FDA approval:**
    Yes. Date of last inspection: April 15, 1988.
14. **Accreditation:**
15. **Communication capabilities:**
    Telephone number: (201) 762-0045
    Telex number: 762 4685
    Computer availability: 4.
16. **Additional data:**
    Total Organic Halogens.
17. **Key contact:**
    C. N. Mangieri

1. **Name of company:**
   Southwest Research Institute
2. **Address:**
   6220 Culebra Road
   Drawer 28510
   San Antonio, Texas 78284-2900
3. **Corporate officers:**
   Martin Goland, President
   H. N. Abramson, Executive Vice President
   M. E. Gates, Executive Vice President
   J. D. Bates, Secretary/Treasurer
4. **Services offered:**
   Analytical, priority pollutants, CLP, drinking water, pesticides, drugs, toxicology testing, toxicology research, environmental analyses.
5. **Tests performed:**
   Environmental, biota, acute, subchronic, chronic toxicity; smoke toxicity tests (New York State and Yew York City requirements); smoke toxicity tests (NBS, radiant heat); smoke toxicity—analytical (for evaluation of major toxicants generated by materials).
6. **Products tested:**
   Pesticides, industrial products, pharmaceuticals, plastics (smoke toxicity testing).
7. **Special expertise:**
   Pharmacokinetics, air monitoring, combustion toxicology, industrial/hygiene, cardiopulmonary, behavioral/neurobehavioral.
8. **Size of lab:**
   Environmental sciences: 20,000 square feet; fire technology: 15,000 square feet.
9. **Equipment available:**
   GC, HPLC, GC/MS, NMR, IR, scintillation counter, FTIR, SFC/MS, IC, ICAP, AA.
10. **Key personnel:**
    Harold L. Kaplan, Ph.D.
    Robert E. Lyle, V.P.
    Gordon E. Hartzell, Ph.D.
    Walter G. Switzer
    Howard W. Stacy
    Donald E. Johnson, Ph.D.
    Donald J. Mangold, Ph.D.
    Richard Geary, Ph.D.
11. **Staff:**
    110.
12. **GLP compliance:**
    Yes. Date of last inspection: May 19, 1986.
13. **FDA approval:**
    Yes. Date of last inspection: August 24, 1984.
14. **Accreditation:**
    EPA, Texas, New York, Oklahoma certification; AAALAC, AALAS, USDA licensed.
15. **Communication capabilities:**
    Telephone number: (512) 522-2171
    Telefax number: (512) 522-3042
    Telex number: 244846
16. **Additional data:**
    Total Organic Halogens. Inhalation toxicology research; respiratory, behavioral and neurobehavioral; toxicology research with rodents and nonhuman primates; combustion toxicology research with rodents and nonhuman primates.
17. **Key contact:**
    Donald E. Johnson, Ph.D.

LISTING OF TOXICOLOGY LABORATORIES

1. **Name of company:**
   Springborn Life Sciences, Inc., Environmental Toxicology & Chemistry Division
2. **Address:**
   790 Main Street
   Wareham, Massachusetts 02543
3. **Corporate officers:**
   Robert B. Foster, Laboratory Director
   Ronald J. Breteler, Ph.D., Director, Program Development
   Jeffrey M. Giddings, Ph.D., Manager, Ecological Programs
   Paul H. Fackler, Ph.D., Manager, Analytical Programs
   Donald C. Surprenant, Manager, Ecotoxicology Programs
   John P. Martinson, Manager, Environmental Fate Programs
4. **Services offered:**
   Ecotoxicological testing; analytical services, toxicity reduction evaluations, field ecological programs, terrestrial ecotoxicological testing, environmental fate programs, consulting programs.
5. **Tests performed:**
   Acute toxicity; chronic toxicity; full lifecycle toxicity evaluations in fish and invertebrates; freshwater and salt water mediums; static, semiflow-through, and flow-through conditions; sediment toxicity; microbial toxicology programs; all routing freshwater and saltwater testing species; phytotoxicity; earthworm toxicity; plant toxicity; field dissipation studies; crop residue programs; metabolism identification programs; mesocosm and microcosm programs; metabolism programs; bioconcentration studies.
6. **Products tested:**
   Pesticides, microbial pest control agents, industrial chemicals, pharmaceuticals, animal health drugs, contaminated soils (Superfund sites), effluents and other waste products.
7. **Special expertise:**
   Ecological effects evaluations; environmental fate and ecotoxicological evaluations; toxicity testing, mesocosm designs, and study conduct; fish and invertebrate lifecycle studies with freshwater and marine organisms; invertebrate taxonomy; analytical method development; metabolism identifications.
8. **Size of lab:**
   40,000 square feet.
9. **Equipment available:**
   GC with ECD, dual-flame ionization, FID and NPD detectors; HPLC; liquid scintillation detectors and sample oxidizers; TOC; autoanalyzer; tissue oxidizer; AA; miscellaneous field sampling equipment, including sampling boats.
10. **Key personnel:**
    R. Foster, Laboratory Director   D. Surprenant, Toxicology
    W. Conroy, Quality Assurance   P. Flackler, Ph.D., Analytical Environmental Chemistry
    K. Grandy, Quality Assurance   J. Giddings, Ph.D., Ecological Programs
11. **Staff:**
    Toxicological technical support: 25, analysts: 26, chemical and fate technical support: 30, ecotoxicologists: 49.
12. **GLP compliance:**
    Yes. Date of last inspection: December 17–19, 1986.
13. **FDA approval:**
    Recognized by FDA as a GLP laboratory, FDA has, to date, not audited the lab for GLP compliance in the ecotoxological and environmental fate areas.
14. **Accreditation:**
    SETAC, ASTM.
15. **Communication capabilities:**
    Telephone number: (508) 295-2550
    Telefax number: (508) 296-8107
    Telex number: 4436041
    Computer availability: no on-line available.
16. **Additional data:**
    Total Organic Halogens.
17. **Key contact:**
    Ronald J. Breteler, Ph.D.

1. **Name of company:**
   Springborn Life Sciences, Inc., Toxicology & Human Safety Division
2. **Address:**
   553 North Broadway
   Spencerville, Ohio 45887
3. **Corporate officers:**
   Dean E. Rodwell, Laboratory Director
   James T. F. Liao, D.V.M., Ph.D., Director of Toxicology
   Kevin G. Michlewicz, Ph.D., Director, Analytical Chemistry
   Michael D. Mercieca, Study Director
   Joseph C. Siglin, Study Director
4. **Services offered:**
   Acute, subchronic, and chronic battery of mammalian toxicology studies and reproductive evaluations of product safety.
5. **Tests performed:**
   Protocols are available for the acute battery of studies including the hypersensitivity and photosensitivity in guinea pigs and inhalation: subchronic and reproductive toxicity. Protocols are specific for EPA/FIFRA guidelines, EPA/TSCA test standards, FDA environmental assessment guidelines, and guidelines for OECD, Japan, and the United Kingdom.
6. **Products tested:**
   Pesticides, industrial chemicals, pharmaceuticals, animal health drugs.
7. **Special expertise:**
   Traditional areas of expertise include guinea pig delayed-contact hypersensitivity and skin photosensitivity studies, acute inhalation, and subchronic and chronic toxicity testing with special emphasis on anticancer drugs. During the past 4 years, added specialty includes teratology and reproductive performance testing.
8. **Size of lab:**
   24,000 square feet.
9. **Equipment available:**
   GC with ECD, dual-flame ionization, FID, and NPD detectors; HPLC with RAM and fluorescent detectors; IR; variable wavelength spectrophotometer; liquid scintillation detectors and simple oxidizers; TOC; autoanalyzer; tissue oxidizer; AA; ECG; all necessary equipment for necropsy and microscopic pathology.
10. **Key personnel:**
    Anita Bosau, Director of Quality Assurance
    Robert Geil, D.V.M., Pathologist, Diplomate A.C.V.P.
    William Collins, D.V.M., Ph.D., Veterinarian and Microscopic Pathologist
    Kevin Michlewicz, Ph.D., Analytical Chemist
    Gabriela Adam, Ph.D., Pathologist/Toxicologist
11. **Staff:**
    Toxicological technical support: 27, analysts: 5, chemical technical support: 4, toxicologists: 35.
12. **GLP compliance:**
    Strict compliance with EPA and FDA GLP.
13. **FDA approval:**
    Underwent five inspections without significant findings.
14. **Accreditation:**
    AAALAC.
15. **Communication capabilities:**
    Telephone number: (419) 647-4196
    Telefax number: (419) 647-6560
    Telex number: 4436041
    Computer availability: On-line available.
16. **Additional data:**
    Total Organic Halogens.
17. **Key contact:**
    Ronald J. Breteler, Ph.D.

LISTING OF TOXICOLOGY LABORATORIES

1. **Name of company:**
   SRI International, Life Sciences Division
2. **Address:**
   333 Ravenswood Avenue
   Menlo Park, California 94025-3493
3. **Corporate officers:**
   William F. Miller, President/CEO
   Paul J. Jorgensen, Executive Vice President/COO
   George R. Abrahamson, Senior Vice President, Sciences Group
   Myron DuBain, Chairman of the Board
4. **Services offered:**
   Research and consulting in an array of disciplines offering toxicology testing, metabolism studies, environmental studies, and analytical services.
5. **Tests performed:**
   Mammalian toxicology: acute, subchronic, chronic; genetic toxicology; reproductive and developmental toxicology; subcellular toxicology; cellular toxicology and pharmacology; neurobehavioral toxicology; health-risk analysis.
6. **Products tested:**
   Pharmaceuticals, medical devices, industrial chemicals, petroleum products, agrochemicals, pesticides, biotechnology products, consumer products.
7. **Special expertise:**
   Excellent experience and reputation in performing routine toxicology testing. With our research orientation and expert scientists, we continue to maintain a leading edge in new areas of toxicology research developments.
8. **Size of lab:**
   130,000 square feet.
9. **Equipment available:**
   LABCAT computer system, EM, ARTEK computer system, flow cytometer.
10. **Key personnel:**
    Robert M. Sutherland, Ph.D., Vice President, Life Sciences Division
    Jon C. Mirsalis, Ph.D., DABT, Director, Cellular and Genetic Toxicology
    James R. Hill, D.V.M., Director, Pathology
    Gordon T. Pryor, Ph.D., Director, Neurosciences Department
11. **Staff:**
    Toxicologists and senior scientists: 19, scientists: 23, technicians: 14.
12. **GLP compliance:**
    Yes. Date of last inspection: August 1987.
13. **FDA approval:**
    Yes. Date of last inspection: August 1987.
14. **Accreditation:**
    AAALAC.
15. **Communication capabilities:**
    Telephone number: (415) 859-3000
    Telefax number: (415) 859-3153
    Telex number: 671-7705
    Computer availability: TYMET.
16. **Additional data:**
    SRI has a Human Cell Culture Facility, which provides a unique capability to conduct comparative metabolism and toxicity studies with normal human tissue. SRI offers research and consulting in medicinal chemistry, molecular biology, analytical chemistry, agricultural biology, radiolabeled synthesis, and immunobiology. We can design a unified multidisciplinary research program to suit a client's specific needs.
17. **Key contact:**
    Erica K. Loh

1. **Name of company:**
   Stillmeadow, Inc.
2. **Address:**
   9525 Town Park Drive
   Houston, Texas 77036
3. **Corporate officers:**
   Robert J. Sobol, President
   Elizabeth J. Sobol, Secretary/Treasurer
4. **Services offered:**
   Toxicology, microbiology, biomedical device evaluation.
5. **Tests performed:**
   Acute oral toxicity, mutagenicity, primary skin irritation, teratology, USP testing, rabbit eye irritation, hen neurotoxicity, acute inhalation, indoor air quality testing, acute dermal toxicity, microbiological studies, guinea pig sensitization, sterility testing, subchronic studies, USP systemic injection test, USP intracutaneous test, USP intramuscular implant, USP transfusion and infusion assembly test, particulate counts, tissue culture agar overlay, inhibition of cell growth, extract cytotoxicity, hemolysis.
6. **Products tested:**
   Pesticides, cosmetics, pharmaceuticals, industrial chemicals, plastics, medical devices.
7. **Special expertise:**
8. **Size of lab:**
   15,000 square feet.
9. **Equipment available:**
   HPLC, GC, spectrophotometer, hemotology, and clinical chemistry equipment.
10. **Key personnel:**
    Janice O. Kuhn, Ph.D., Study Director
    Mark S. Holbert, B.S., Study Director
    Elizabeth J. Sobol, B.A., B.S., Director, Data Services
    James F. Gregory, M.S., Director of Quality Assurance
    Robert E. Faith, D.V.M., Ph.D., Consulting Veterinarian
    Robert E. Baughn, Ph.D., Consulting Microbiologist
    Stephen B. Harris, Ph.D., Consulting Teratologist
    Larry P. Jones, D.V.M., Consulting Veterinary Pathologist
11. **Staff:**
    Toxicologists: 4, consultants: 5, technicians: 5, support personnel: 9, quality assurance: 2.
12. **GLP compliance:**
    Yes. Date of last inspection: December 1988.
13. **FDA approval:**
    Yes. Date of last inspection: December 1988.
14. **Accreditation:**
    AAALAC.
15. **Communication capabilities:**
    Telephone number: (713) 776-8828
    Telefax number: (713) 271-9779
16. **Additional data:**
17. **Key contact:**
    Robert J. Sobol

LISTING OF TOXICOLOGY LABORATORIES

1. **Name of company:**
   Structure Probe, Inc.
2. **Address:**
   Box 656
   West Chester, Pennsylvania 19381-0656
3. **Corporate officers:**
   Charles A. Garber, Ph.D., President/CEO
4. **Services offered:**
   Independent laboratory offering scanning and transmission electron and light microscopy services; other major capabilities: quantitative image analysis, ion milling, cryo-ultramicrotomy.
5. **Tests performed:**
   Asbestos in air via TEM (Ahera procedure), asbestos in bulk samples via PLM/DS (NIOSH procedure), asbestos in water (EPA Procedure).
6. **Products tested:**
   Plastics, polymer coatings, elastomers, metals, cosmetics and toiletries, composite materials, ceramics, semiconductor devices, life sciences, pharmaceuticals.
7. **Special expertise:**
   SEM, TEM, PLM, x-ray diffraction, polymer physics, asbestos detection, polymer failure analysis, rubber modified polymers, polymer coatings/ceramics, electronics/metals.
8. **Size of lab:**
   10,000 square feet.
9. **Equipment available:**
   TEM, SEM, ultramicrotomes, ion milling device, vacuum evaporators, sputter coaters, plasma etchers, light microscope, image analysis systems.
10. **Key personnel:**
    Kim Royer, Operations Manager
    Gene Rodek, SEM Services
    Judy Cosby, TEM Services
11. **Staff:**
    10.
12. **GLP compliance:**
    N/A.
13. **FDA approval:**
    N/A.
14. **Accreditation:**
    AAALA—in microscopy subgroup. AIHA (in process), NIOSH PAT program for bulk asbestos, ACIL, SAMPE, Electron Microscopy Society of America.
15. **Communication capabilities:**
    Telephone number: (215) 436-5400, (800) 2424-SPI
    Telefax number: (215) 436-5755
    Telex number: 835367
16. **Additional data:**
    Analytical. Affiliated Labs: Structure Probe, Inc., Fairfield, Connecticut; Structure Probe, Inc., Copiague, New York; Structure Probe, Inc., Metuchen, New Jersey.
17. **Key contact:**
    Charles A. Garber, Ph. D.

1. **Name of company:**
   Structure Probe, Inc.
2. **Address:**
   63 Unquowa Road
   Fairfield, Connecticut 06430
3. **Corporate officers:**
   Charles A. Garber, Ph.D., President/CEO
4. **Services offered:**
   Independent laboratory offering scanning electron microscopy and energy dispersive x-ray analysis services.
5. **Tests performed:**
   Metallurgical failure analysis, circuit board failure analysis, composite materials failure analysis, cosmetics/toiletries claims documentation testing.
6. **Products tested:**
   Plastics, polymer coatings, elastomers, metals, cosmetics and toiletries, composite materials, ceramics, semiconductor devices, life sciences, pharmaceuticals.
7. **Special expertise:**
   Electron microscopy SEM/TEM/EDS, surface analysis AUGER/XPS/SIMS/LIMS/ISS, x-ray diffraction.
8. **Size of lab:**
   2,000 square feet.
9. **Equipment available:**
   SEM, EDP.
10. **Key personnel:**
    A. W. Blackwood, Ph.D., Laboratory Manager/Director
    Ronald Bucari, Microscopist
11. **Staff:**
    2.
12. **GLP compliance:**
    N/A.
13. **FDA approval:**
    N/A.
14. **Accreditation:**
    AALA—microscopy subgroup.
15. **Communication capabilities:**
    Telephone number: (203) 254-0000
    Telefax number: (203) 254-2262
16. **Additional data:**
    See information for West Chester, Pennsylvania office.
17. **Key contact:**
    Charles A. Garber, Ph.D.

LISTING OF TOXICOLOGY LABORATORIES

1. **Name of company:**
   Structure Probe, Inc.
2. **Address:**
   1015 Merrick Road
   Copiague, New York 11726
3. **Corporate officers:**
   Charles A. Garber, Ph.D., President/CEO
4. **Services offered:**
   Independent laboratory offering scanning electron microscopy and energy dispersive x-ray spectroscopy services.
5. **Tests performed:**
   See information for Fairfield, Connecticut office.
6. **Products tested:**
   See information for Fairfield, Connecticut office.
7. **Special expertise:**
   See information for Fairfield, Connecticut office.
8. **Size of lab:**
   1,500 square feet.
9. **Equipment available:**
   SEM, EDP.
10. **Key personnel:**
    A. W. Blackwood, Ph.D., Laboratory Manager
11. **Staff:**
    2.
12. **GLP compliance:**
    N/A.
13. **FDA approval:**
    N/A.
14. **Accreditation:**
    ACIL.
15. **Communication capabilities:**
    Telephone number: (516) 789-0100
16. **Additional data:**
    See information for West Chester, Pennsylvania office.
17. **Key contact:**
    Charles A. Garber, Ph.D.

1. **Name of company:**
   Structure Probe, Inc.
2. **Address:**
   230 Forrest Street
   Metuchen, New Jersey 08840
3. **Corporate officers:**
   Charles A. Garber, Ph.D., President/CEO
4. **Services offered:**
   Independent laboratory offering surface analysis and electron microscopy services, including auger electron spectroscopy, x-ray photoelectron spectroscopy, scanning electronic microscopy, electron probe microanalysis.
5. **Tests performed:**
   Metallurgical failure analysis, adhesion failure analysis.
6. **Products tested:**
   Plastics, polymer coatings, elastomers, metals, cosmetics and toiletries, composite materials, ceramics, semiconductor devices, life sciences, and pharmaceuticals.
7. **Special expertise:**
   SEM, AES, XPS, EPM, EDS; polymer physics, asbestos detection, polymer failure analysis, rubber modified polymers, polymer coatings, ceramics/electronics/metals.
8. **Size of lab:**
   3,000 square feet.
9. **Equipment available:**
   SEM, EDS, AES, XPS, cryotransfer SEM, metallographic preparation facilities.
10. **Key personnel:**
    J. S. Duerr, Ph.D., P.E., Metallurgy
    Andrew Hirt, Physics
    Nora Ross, Ceramics
    Myra James, Biology
    Abdul Jabbar, Surface Physics
11. **Staff:**
    7.
12. **GLP compliance:**
    N/A.
13. **FDA approval:**
    N/A.
14. **Accreditation:**
    American Association for Laboratory Accreditation (AALA)—microscopy subgroup.
15. **Communication capabilities:**
    Telephone number: (201) 549-9350
    Telefax number: (201) 549-9356
16. **Additional data:**
    Analytical. See information for West Chester, Pennsylvania office.
17. **Key contact:**
    Charles A. Garber, Ph.D.

1. **Name of company:**
   Syracuse Research Corporation
2. **Address:**
   Merrill Lane
   Syracuse, New York 13210-4080
3. **Corporate officers:**
   Dr. Kenneth A. Kun, President/CEO
   Dr. Adam Kozma, Vice President/Director of Engineering
   Paul J. Doherty, Director of Finance/CFO
   Cheryl S. Saunders, Secretary/Director of Personnel
   Dr. Robert A. Herman, Director of New Business Development
4. **Services offered:**
   Analytical services, toxicology testing, industrial and environmental consulting, environmental fate studies.
5. **Tests performed:**
   Static acute, dynamic (flow-through), static renewal, freshwater and marine species, algae.
6. **Products tested:**
   Environmental samples, food, industrial chemicals, pharmaceuticals.
7. **Special expertise:**
   GC/MS, GC, HPLC, aquatic toxicology.
8. **Size of lab:**
   7,785 square feet.
9. **Equipment available:**
   GC/MS, GC, HPLC, AA.
10. **Key personnel:**
    Alison E. Carter, Ph.D.
    Francis G. Doherty, Ph.D.
    Whei-Chu Pan, M.S.
    Craig R. Turner, B.A.
    Susan L. Pratt, B.S.
    Lenore A. Gooden, B.S.
    Hugh M. Guider, B.S.
    David J. Prichard, A.A.S.
11. **Staff:**
12. **GLP compliance:**
    N/A.
13. **FDA approval:**
    Certified for the bacteriological examination of single-service containers and closures. Date of last inspection: February 1988.
14. **Accreditation:**
    New York State Department of Health certified for the analysis of drinking water, wastewater, and solid waste.
15. **Communication capabilities:**
    Telephone number: (315) 425-5100
    Telefax number: (315) 425-1339
    Computer availability: Central VAX.
16. **Additional data:**
    Total Organic Halogens.
17. **Key contact:**
    Dr. Alison E. Carter

1. **Name of company:**
   Thermo Analytical, Inc.
2. **Address:**
   160 Taylor Street
   Monrovia, California 91016
3. **Corporate officers:**
   Roger Herd, President
   Michelle Miller, Vice President/General Manager
   John D. McCarthy, Business Development Manager
4. **Services offered:**
   Analytical, chemistry, quality control, field sampling.
5. **Tests performed:**
   EPA methods—hazardous waste and wastewater; NIOSH methods—industrial hygiene.
6. **Products tested:**
   Hazardous waste, soil contaminants; priority pollutants; water; wastewater; landfill gas, air, fuel gas.
7. **Special expertise:**
   Component analysis of organic/inorganic mixtures using various techniques—FTIR, NMR, pyrolysis; plastic analysis.
8. **Size of lab:**
   8,000 square feet.
9. **Equipment available:**
   GC, AA, UV/VIS, ICP, GC/MS, FTIR.
10. **Key personnel:**
    Shirley Kirby, Metals Supervisor
    Dennis Wells, Operations Manager
    Dave Kohlenberger, Quality Control Manager
11. **Staff:**
    19.
12. **GLP compliance:**
13. **FDA approval:**
14. **Accreditation:**
    California Department of Health Hazardous Waste Certification, USEPA-DAT Asbestos Certification, NIOSH-AIHA Accreditation (industrial hygiene).
15. **Communication capabilities:**
    Telephone number: (818) 357-3247
    Telefax number: (818) 359-5036
    Computer availability: Modem on IBM-AT.
16. **Additional data:**
    Consulting chemists.
17. **Key contact:**
    John D. McCarthy

LISTING OF TOXICOLOGY LABORATORIES

1. **Name of company:**
   Toxikon Corporation
2. **Address:**
   225 Wildwood Avenue
   Woburn, Massachusetts 01801
3. **Corporate officers:**
   Laxman S. Desai, D.S.C., President and Director
   Herman S. Lilja, Ph. D., Vice President, Director of Toxicology and Operations
4. **Services offered:**
   Toxicology testing, analytical chemistry services.
5. **Tests performed:**
   Toxicology—acute, subchronic, chronic, gene-tox, in vivo/in vitro rats, mice, rabbits, guinea pigs, hamsters; cells in culture, bacteria.
6. **Products tested:**
   Pesticides, medical devices, industrial chemicals, biotechnology products, cosmetics, pharmaceuticals.
7. **Special expertise:**
   Pharmacokinetics, metabolite identification, drug distribution, environmental fate, residue analysis.
8. **Size of lab:**
   23,000 square feet.
9. **Equipment available:**
   GC/MS, ICP, AA, GC, IR, UV/VIS, HPLC, rapid flow analyzer, spectrofluorometer.
10. **Key personnel:**
    Nancy DiGiulio, B.S., Director, Quality Assurance
    Paul M. Lexberg, B.S., Coordinator, Analytical Chemistry
    Amy L. Austin, B.S., Manager, Animal Toxicology
    Wubishet Tilahun, B.S., Supervisor, Organic Chemistry
    Mark A. Devlin, B.S., Supervisor, Inorganic Chemistry
11. **Staff:**
    Analytical chemistry: 5, computer data management: 3, technicians: 7, quality assurance: 2, toxicologists: 4.
12. **GLP compliance:**
    And GMP: FDA and EPA, Date of last inspection: March 1988.
13. **FDA approval:**
    Registration. Date of last inspection: March 1988.
14. **Accreditation:**
    AAALAC, VSDA, EPA, DEQE.
15. **Communication capabilities:**
    Telephone number: (617) 933-6903
    Telefax number: (617) 933-9196
    Telex: 924-441
    Computer Availability: In-house computer network in place, on-line service by September 1988.
16. **Additional data:**
    Total Organic Halogens.
17. **Key contact:**
    Herman S. Lilja, Ph.D.

1. **Name of company:**
   TPS, Inc.
2. **Address:**
   10424 Middle Mt. Vernon Road
   Mt. Vernon, Indiana 47620
3. **Corporate officers:**
   Dr. James Botta, Jr.
4. **Services offered:**
   Toxicology testing in mammalian species.
5. **Tests performed:**
   Acute to chronic in rats, mice, dogs, guinea pigs, rabbits, monkeys, swine.
6. **Products tested:**
   All.
7. **Special expertise:**
   Pharmacokinetics, reproduction, dermal penetration.
8. **Size of lab:**
   40,000 square feet.
9. **Equipment available:**
   HPLC, GC.
10. **Key personnel:**
    James Botta, Jr., D.V.M., Ph.D., Scientific Director
    John Wedig, Ph.D., DABT, Director of Toxicology
    D. C. Larson, Ph.D., Director of Nutritional Safety
11. **Staff:**
    Technicians: 25.
12. **GLP compliance:**
    Yes. Date of last inspection: March 1988.
13. **FDA approval:**
    Yes. Date of last inspection: March 1988.
14. **Accreditation:**
15. **Communication capabilities:**
    Telephone number: (812) 985-5900
    Telefax number: (812) 985-3403
    Computer availability: PDP 11/73 Digital CPU with 160MB on-line storage; completely computerized facility.
16. **Additional data:**
    Total Organic Halogens.
17. **Key contact:**
    Dr. John Wedig

1. **Name of company:**
   Roy F. Weston, Inc.
2. **Address:**
   Weston Way
   West Chester, Pennsylvania 19380
3. **Corporate officers:**
   Roy F. Weston, Chairman of the Board
   Thomas M. Swoyer, President
   William S. Gaither, Vice Chairman (RFW)/President (NCI)
   A. Frederick Thompson, Executive Vice President/Assistant Secretary
   Jack C. Newell, Executive Vice President
   M. Salah Abdelhamid, Vice President
   George J. Anastos, Vice President
4. **Services offered:**
   Analytical chemistry in support of environmental management.
5. **Tests performed:**
6. **Products tested:**
   Air, water, wastewater, soil, sediment, sludge, waste for an extensive array of organic and inorganic parameters.
7. **Special expertise:**
   Participate in the EPA interlaboratory performance program for water (WS) and wastewater (WP).
8. **Size of lab:**
   All 3 facilities: 70,000 square feet.
9. **Equipment available:**
   GC, GC/MS, AA, ICP, HPLC, UV/VIS, TOC, TOX, IC.
10. **Key personnel:**
    C. P. Nulton, Ph.D., Lionville Laboratory Manager
    N. W. Flynn, Project Director
    R. A. Elliott, Stockton Laboratory Manager
    J. M. Taylor, Field Laboratory Manager
    J. P. Boudreau, Gulf Coast Laboratory Technical Director/Vice President
11. **Staff:**
    Ph.D.: 6, M.S.: 12, B.S.: 99, A.A.S.: 19, total personnel: 200.
12. **GLP compliance:**
    N/A.
13. **FDA approval:**
    N/A. NRC: Lionville licensed for low-level; Stockton: applied for license.
14. **Accreditation:**
    USEPA (CLP), U.S. Army Corps of Engineers, USATHAMA, U.S. Air Force, NRC, certified in 27 of 34 states that have programs.
15. **Communication capabilities:**
    Telephone number: (215) 524-7360   (215) 363-8622
    Telefax number: (215) 524-7503   (215) 524-1723
16. **Additional data:**
    Analytical.
17. **Key contact:**
    Jack C. Newell, P.E.

1. **Name of company:**
   White Eagle Toxicology Laboratories
2. **Address:**
   2003 Lower State Road
   Doylestown, Pennsylvania 18901
3. **Corporate officers:**
   Abbott S. D'Ver, President
   William J. Ehrhart, Vice President
4. **Services offered:**
   Neonatal, pediatric, and adult toxicology; teratology; fertility.
5. **Tests performed:**
   Acute; subchronic; chronic; teratology; efficacy on dogs, cats, rabbits, guinea pigs, rats, mice.
6. **Products tested:**
   Pharmaceuticals, industrial chemicals, pesticides.
7. **Special expertise:**
   Neonatal toxicology, developmental toxicology.
8. **Size of lab:**
   20,000 square feet.
9. **Equipment available:**
10. **Key personnel:**
    Edward Schwartz, D.V.M., Ph.D.
    Abbott S. D'Ver, D.V.M.
    Stan Sekelewski
    William J. Ehrhart
11. **Staff:**
    10.
12. **GLP compliance:**
    Continually.
13. **FDA approval:**
    Yes. Date of last inspection: 1988.
14. **Accreditation:**
15. **Communication capabilities:**
    Telephone number: (215) 348-3868
    Telefax number: (215) 348-5081
16. **Additional data:**
17. **Key contact:**
    Abbott S. D'Ver, D.V.M.

LISTING OF TOXICOLOGY LABORATORIES

1. **Name of company:**
   Wildlife International Ltd.
2. **Address:**
   305 Commerce Drive
   Easton, Maryland 21601
3. **Corporate officers:**
   Curt Hutchinson
   Mark Jaber
4. **Services offered:**
   Toxicological, analytical. avian toxicology, aquatic mesocosm, aquatic field programs, terrestrial field programs, runoff and dissipation studies, ecotoxicology, aquatic toxicology, analytical programs.
5. **Tests performed:**
   Acute avian toxicity, avian reproduction, poultry residue and metabolism, nontarget insects, agricultural-related programs.
6. **Products tested:**
   Pesticides, pharmaceuticals, industrial chemicals, industrial effluents.
7. **Special expertise:**
   Ecotoxicity.
8. **Size of lab:**
   40,000 square feet.
9. **Equipment available:**
   All necessary equipment for conducting laboratory and field programs.
10. **Key personnel:**
    Alan Hosmer
    Hank Krueger
    Joann Beavers
    Cindy Driscoll
11. **Staff:**
    Avian toxicologists: 12, terrestrial ecologists: 20, aquatic toxicologists: 25, seasonal and temporary employees: 47.
12. **GLP compliance:**
    Yes. Date of last inspection: EPA, January 1989.
13. **FDA approval:**
    Facilities audit passed. Date of last inspection: August 7–8, 1984.
14. **Accreditation:**
15. **Communication capabilities:**
    Telephone number: (301) 822-8600
    Telefax number: (301) 822-0632
    Telex Number: 87414 LANDESTN
16. **Additional data:**
    Total Organic Halogens. Conducts field programs in 15 states, has ability to rapidly mobilize for large-scale programs, and has reputation for meeting stringent time tables with high-quality studies.
17. **Key contact:**
    Curt Hutchinson

1. **Name of company:**
   Wizard Laboratories, Inc.
2. **Address:**
   1362 Monarch Lane
   Davis, California 95616
3. **Corporate officers:**
   Alan J. Brattesani, President
4. **Services offered:**
   Custom synthesis of C-14 radiolabeled compounds.
5. **Tests performed:**
6. **Products tested:**
7. **Special expertise:**
   Radiolabeled synthesis.
8. **Size of lab:**
9. **Equipment available:**
   Radio HPLC, radio TLC scanner, GC, scintillation counter.
10. **Key personnel:**
11. **Staff:**
12. **GLP compliance:**
    N/A.
13. **FDA approval:**
    N/A. Note: NRC—California state radioactive materials license, current.
14. **Accreditation:**
15. **Communication capabilities:**
    Telephone number: (916) 753-6700
16. **Additional data:**
    Special services.
17. **Key contact:**
    Alan J. Brattesani

Part II

Appendices

Appendix A

Summary List of Laboratory Contacts

Name of company	Street name	Box no.	City
American Health Foundation	1 Dana Rd.		Valhalla
American Radiolabeled Chemicals, Inc.	11612 Bowling Green Dr.		St. Louis
Amersham Corp.	2636 S. Clearbrook Dr.		Arlington Heights
Ana-Lab Corp.	2600 Dudley Rd.		Kilgore
Analytical Bio-Chemistry Laboratories, Inc.	7200 East ABC Lane		Columbia
Ani Lytics, Inc.	360 Christopher Ave.		Gaithersburg
Applied Genetics, Inc.	1335 Gateway Dr. #2001		Melbourne
Argus Research Laboratories, Inc.	935 Horsham Rd.		Horsham
Arthur D. Little Chemical & Life Sciences Section	25 Acorn Park		Cambridge
Battelle Memorial Institute	30 Memorial Drive 505 King Ave.		Columbus
Bio-Life Associates Ltd.	Route 3	156	Neillsville
Bio/dynamics, Inc.	Mettlers Rd.	2360	East Millstone
Biological Test Center	2525 McGaw Ave.		Irvine
BIOMED, Inc.	1720 130th Ave., N.E.		Bellevue
Bionetics Research, Inc.	5516 Nicholson Lane		Kensington
Biospherics, Inc.	12051 Indian Creek Court		Beltsville
Bushy Run Research Center	R.D. 4, Mellon Rd.		Export
Chem Services, Inc.	660 Tower Lane	3108	West Chester
Chemsyn Science Laboratories (Member of the Specialty Materials Division, Eagle-Picher Industries, Inc.)	13605 W. 96th Terrace		Lenexa
Colorado Histo-Prep, Inc.		8644	Fort Collins
Comparative Toxicology Laboratories VCS Kansas State University			Manhattan
Cosmopolitan Safety Evaluation, Inc.	Statesville Quarry Rd.	71	Lafayette
Dawson Research Corp.		620666	Orlando
Ecology & Environment, Inc., Analytical Services Center	4285 Genesee St.		Buffalo
ECS/Normandeau		1393	Aiken
Education & Research Foundation, Inc.	2602 Langhorne Rd.		Lynchburg
Enseco, Inc.	Doaks Lane at Little Harbor		Marblehead
Enviroscan, Inc.	303 W. Military Rd.		Rothschild
Essex Testing	799 Bloomfield Ave.		Verona
Food and Drug Research Laboratories Division of Enviro/Analysis Corp.	Rt. 17C	107	Waverly
Harris Laboratories, Inc.	624 Peach St.	80837	Lincoln
Hazards Research Corp.	200 East Main St.		Rockaway
Hazelton Laboratories America, Inc.	3301 Kinsman Blvd.		Madison
Hunter Environmental Services, Inc.	900 Osceola Dr.	1703	West Palm Beach Gainesville
IIT Research Institute	10 W. 35th St.		Chicago
ImmuQuest Laboratories, Inc.	13 Taft Court, Suite 200		Rockville
Kemron Environmental Services, Inc.	755 New York Ave.		Huntington
La Rocca Science Laboratories, Inc.	1 Nell Court		Dumont
Laboratory Research Enterprises, Inc.	6321 South 6th St.		Kalamazoo
Lancaster Laboratories, Inc.	2425 New Holland Pike		Lancaster
Langston Laboratories, Inc.	2006 West 103rd Terrace		Leawood
Leberco Testing, Inc.	123 Hawthorne St.		Roselle Park
Lycott Environmental Research, Inc.	600 Charlton St.		Southbridge
M. B. Research Laboratories, Inc.	Steinberg & Wentz Rds.	178	Spinnerstown
Malcolm Pirnie, Inc.	2 Corporate Park Dr. 100 Grasslands Rd.		White Plains Elmsford
Microbiological Associates, Inc.	9900 Blackwell Rd.		Rockville
Midwest Research Institute	425 Volker Blvd.		Kansas City
Nebraska Testing Corp.	4453 South 67th St.		Omaha
North American Science Associates, Inc.	2261 Tracy St. 9 Morgan		Northwood Irvine
Northview Pacific Laboratories, Inc.	2800 Seventh St. 1880 Holste Rd.		Berkeley Northbrook
Oak Creek Laboratory of Biology, Department of Fisheries & Wildlife	Nash Hall, Room #104 Oregon State University		Corvallis

APPENDIX A: SUMMARY LIST OF LABORATORY CONTACTS   87

State	Zip code	Telephone number	Telefax number	Telex number	Key contact
NY	10595	(914) 592-2600	(914) 592-6317		G. M. Williams, M.D.
MO	63146	(314) 991-4545	(314) 991-4692	9102404101 AMMERADCHEM UQ	Surendra Gupta
IL	60005	(312) 593-6300	(312) 593-1044 or 8236		G. T. Anderson
TX	75662	(214) 984-0551			C. H. Whiteside
MO	65202	(314) 474-8579	(314) 443-9033	821814	James B. Rabenold, Ph.D.
MD	20879	(301) 921-0168	(301) 977-0248		Saroj R. Das, Ph.D.
FL	32901-2619	(407) 768-2048	(407) 984-2890		Dr. Maria H. Lugo
PA	19044	(215) 443-8710	(215) 443-8587		Alan M. Hoberman, Ph.D.
MA	02140	(617) 864-5770	(617) 547-3617	921436	Jack Pulidoro, Ph.D.
OH	43201-2693	(614) 424-5836	(614) 424-5263	24-5454	Dr. Jake Halliday
WI	54456	(715) 743-4557			Dale W. Fletcher
NJ	08875-2360	(201) 873-2550	(201) 873-3992	844597 BIO DYN EMLS	Lynn J. Corrigan
CA	92714	(714) 660-3185	(714) 660-2565	4970362	Paul Mazur, Ph.D.
WA	98005-2203	(206) 882-0448	(206) 882-2678	283803 BIOM	Mark Levine
MD	20895-1078	(301) 881-5600	(301) 984-3608	89-8369	Sue C. Tondreau
MD	12051	(301) 369-3900	(301) 725-4908	898 072	Robert G. Edwards, Ph.D.
PA	15632	(412) 733-5200	(412) 733-4804		William M. Snellings, Ph.D.
PA	19381-3108	(215) 692-3026		510-663-0003	E. Hollenbach
KS	66215	(913) 541-0525 (800) 233-6643	(913) 888-3582	(910) 840-3270	Lisa Bosch
CO	80524	(303) 493-2660			Janet E. Maass
KS	66506	(913) 532-5679			F. W. Oehme
NJ	07848	(201) 383-6253	Restricted to clients		Dr. G. R. Robbins
FL	32862-0666	(407) 851-3110	(407) 851-3110		Dr. Thomas E. Murchison
NY	14225	(716) 631-0360			John Gartner
SC	29802	(803) 652-2206	(803) 652-7428		Kathleen E. Trapp
VA	24501	(804) 847-5695	(804) 846-1707		Bert Mathews
MA	01945	(617) 639-2695	(617) 639-2637		Timothy J. Ward
WI	54474	(715) 359-7226 (800) 338-7226	(715) 355-3219	29-0495	James M. Force
NJ	07044	(201) 857-9541	(201) 857-9662		Harold Schwartz
NY	14892	(607) 565-8131	(607) 565-7420		John A. Biesemeier
NB	68501	(402) 476-2811	(402) 476-7598	230-469-018	Robert P. Marshall
NJ	07866	(201) 627-4560	(201) 627-0015		William J. Cruice
WI	53704	(608) 241-4471	(608) 241-7227	703956 HAZRALMDS UD	Robert E. Conway
FL	33409	(407) 686-7210	(407) 471-5295		Charlene Bowman
FL	32602-1703	(904) 332-3318	(904) 332-0507		G. Scott Ward
IL	60616-3799	(312) 567-4000	(312) 567-4577	282472 IITRI CGO	Dr. James D. Fenters
MD	20850	(301) 762-7193			James L. McCoy
NY	11743	(516) 427-0950	(516) 427-0989		Maureen J. Nimphius
NJ	07628	(201) 384-8509			Rudolph La Rocca
MI	49009-9611	(616) 375-0482	(616) 375-8403		Martin R. Gilman, Ph.D.
PA	17601	(717) 656-2301	(717) 656-2681		Randall H. Guthrie
KS	66206	(913) 341-7800			Don P. Miller
NJ	07204-0206	(201) 245-1933	(201) 245-6253		Dr. E. C. Rothstein
MA	01550	(508) 765-0101			Ron LeBlanc
PA	18968	(215) 536-4110	(215) 536-1816		Oscar M. Moreno, Ph.D.
NY	10602	(914) 347-2970			Jane S. Hughes
NY	10523	(analytical) (914) 347-2974 (aquatic)			
MD	20850	(301) 738-1000	(301) 738-1036	908793	Cheryl Respess
MO	64110	(816) 753-7600	(816) 753-8420	910-771-2128	Margaret Gunde
NB	68117	(402) 331-4453	(402) 331-5961		George C. Phelps
OH	43619	(419) 666-9455	(419) 666-2954	810-442-1732	Suellen R. Romick
CA	92718	(714) 951-3110	(714) 951-3280	628-29106	
		(415) 548-8340	(415) 548-5425		Joanne Spalding
CA	94710				
IL	60062	(503) 754-4531	(503) 754-2400	510-596-0682	Lawrence R. Curtis
OR	97331-3803			OSU COVS	

DIRECTORY OF TOXICOLOGICAL AND RELATED TESTING LABS

Name of company	Street name	Box no.	City
Product Safety Laboratories	725 Cranbury Rd.		East Brunswick
Recon Systems, Inc.	Route 202 North	460	Three Bridges
Research Triangle Institute	3040 Cornwallis Rd.	12194	Research Triangle Park
Ricerca, Inc.	7528 Auburn Rd.		Painesville
Roy F. Weston, Inc.	Weston Way		West Chester
Savannah Laboratories & Environmental Services, Inc.			
Savannah Div.	5102 LaRoche Ave.		Savannah
Mobile Div.	3707 Cottage Hill Rd.		Mobile
Tallahassee Div.	2820 Industrial Plaza		Tallahassee
Scientific Associates, Inc.	6200 South Lindbergh Blvd.		St. Louis
Serco Laboratories	1931 Westcounty Rd. C-2		St. Paul
Shrader Analytical & Consulting Laboratories, Inc.	3814 Vienwood		Detroit
Skinner & Sherman Laboratories, Inc.	300 Second Ave.	521	Waltham
South Mountain Laboratories, Inc.	380 Lackawanna Place		South Orange
Southern Research Institute	2000 Ninth Ave., South	55305	Birmingham
Southwest Research Institute	6220 Culebra Road	Drawer 28510	San Antonio
Springborn Life Sciences, Inc., Toxicology & Human Safety Division	553 North Broadway		Spencerville
	790 Main St.		Wareham
SRI International Life Sciences Division	333 Ravenswood Ave.		Menlo Park
Stillmeadow, Inc.	9525 Town Park Dr.		Houston
Structure Probe, Inc.		656	West Chester
	1015 Merrick Rd.		Copiague
	230 Forrest St.		Metuchen
	63 Unquowa Rd.		Fairfield
Syracuse Research Corp.	Merrill Lane		Syracuse
Thermo Analytical, Inc.	160 Taylor St.		Monrovia
Toxikon Corporation	225 Wildworld Ave.		Woburn
TPS, Inc.	10424 Middle Mt. Vernon Rd.		Mt. Vernon
White Eagle Toxicology Laboratories	2003 Lower State Rd.		Doylestown
Wildlife International Ltd.	305 Commerce Dr.		Easton
Wizard Laboratories, Inc.	1362 Monarch Lane		Davis

APPENDIX A: SUMMARY LIST OF LABORATORY CONTACTS

State	Zip code	Telephone number	Telefax number	Telex number	Key contact
NJ	08816	(201) 254-9200	(201) 254-6736		Walter Newman
NJ	08887	(201) 782-5900	(201) 788-0830		Richard F. Toro
NC	27709	(919) 541-6000	(919) 541-6499	802509 (RTI RTPK)	Lisa C. Jeffrey
OH	44077-1000	(216) 357-3300	(216) 354-4415	196212 RICERCA UT	Joseph A. Kohonoski
PA	19380	(215) 524-7360	(215) 524-7503		Jack C. Newell
		(215) 363-8622	(215) 524-1723		
					Janette Davis Long
GA	31404	(912) 354-7858	(912) 352-0165		
AL	36609	(205) 666-6633	(205) 666-6696		
FL	32301	(904) 878-3994	(904) 878-9504		
MO	63123	(314) 487-6776	(314) 487-3991	447664	Robert H. Moulton
MN	55113	(612) 636-7173	(612) 636-7178		Carol A. Kuehn
MI	48208	(313) 894-4440	(313) 894-4489		Marianne L. Schrader
MA	02254	(617) 890-7200	(617) 890-3883		Dr. John P Appleton
NJ	07079	(201) 762-0045		762 4685	C. N. Mangieri
AL	35255-5305	(205) 581-2000			J. Craig Battles
TX	78284-2900	(512) 522-2171	(512) 522-3042	244846	Donald E. Johnson, Ph.D.
OH	45887	(419) 647-4196	(419) 647-6560	4436041	Ronald J. Breteler, Ph.D.
MA	02543	(508) 295-2550	(508) 296-8107	4436041	Ronald J. Breteler, Ph.D.
CA	94025-3493	(415) 859-3000	(415) 859-3153	671-7705	Erica K. Loh
TX	77036	(713) 776-8828	(713) 271-9779		Robert J. Sobol
PA	19381-9656	(215) 436-5400			
		(800) 2424-SPI	(215) 436-5755	835367	Charles A. Garber
NY	11726	(516) 789-0100			Charles A. Garber
NJ	08840	(201) 549-9350	(201) 549-9356		Charles A. Garber
CT	06430	(203) 254-0000	(203) 254-2262		Charles A. Garber
NY	13210-4080	(315) 425-5100	(315) 425-1339		Dr. Alison E. Carter
CA	91016	(818) 357-3247	(818) 359-5036		John D. McCarthy
MA	01801	(617) 933-6903	(617) 933-9196	924-441	Herman S. Lilja, Ph.D.
IN	47620	(812) 985-5900	(812) 985-3403		Dr. John Wedig
PA	18901	(215) 348-3868	(215) 348-5081		Abbott S. D'Ver
MD	21601	(301) 822-8600	(301) 822-0632	87414 LANDESTN	Curt Hutchinson
CA	95616	(916) 753-6700			A. J. Brattesani

Appendix B

List of Studies Conducted and/or Services Provided

ACUTE

Arthur D. Little, Chemical & Life Sciences Section
Battelle Memorial Institute
Bio/dynamics, Inc.
Biological Test Center
BIOMED, Inc.
Bushy Run Research Center
Comparative Toxicology Laboratories, Kansas State University
Cosmopolitan Safety Evaluation, Inc.
Dawson Research Corporation
Enseco, Inc.
Food & Drug Research Laboratories
Hazleton Laboratories America, Inc.
IIT Research Institute
Laboratory Research Enterprises, Inc.
Leberco Testing, Inc.
M. B. Research Laboratories, Inc.
Microbiological Associates, Inc.
Midwest Research Institute
North American Science Associates
Northview Pacific Laboratories, Inc.
Product Safety Laboratories
Ricerca, Inc.
Scientific Associates, Inc.
South Mountain Laboratories, Inc.
Southwest Research Institute
Springborn Life Sciences, Inc., Toxicology & Human Safety Division
Stillmeadow, Inc.
SRI International, Life Sciences Division
Toxikon Corporation
TPS, Inc.
White Eagle Toxicology Laboratories

ANALYTICAL SERVICES

Ana-Lab Corporation
Arthur D. Little, Chemical & Life Sciences Section
Battelle Memorial Institute
Bio/dynamics, Inc.
Biospherics, Inc.
ECS/Normandeau
Ecology & Environment, Inc.
Enviroscan, Inc.
Hazleton Laboratories America, Inc.
Lancaster Laboratories, Inc.
Lycott Environmental Research, Inc.

Malcolm Pirnie, Inc.
Microbiological Associates, Inc.
Midwest Research Institute
Nebraska Testing Corporation
Product Safety Laboratories
Recon Systems, Inc.
Savannah Laboratories & Environmental Services, Inc.
Scientific Associates, Inc.
Serco Laboratories
Shrader Analytical & Consulting Laboratories, Inc.
Skinner & Sherman Laboratories, Inc.
Southern Research Institute
Southwest Research Institute
Stillmeadow, Inc.
Structure Probe, Inc.
Syracuse Research Corporation

AQUATIC

Analytical Bio-Chemistry Laboratories, Inc.
Ecology & Environment, Inc., Analytical Services Center
Enseco, Inc.
Malcolm Pirnie, Inc.
Oak Creek Laboratory of Biology, Department of Fisheries & Wildlife
Wildlife International Ltd.

CHRONIC (INCLUDES CARCINOGENICITY)

American Health Foundation
Arthur D. Little, Chemical & Life Sciences Section
Battelle Memorial Institute
Bio/dynamics, Inc.
BIOMED, Inc.
Bushy Run Research Center
Dawson Research Corporation
Enseco, Inc.
Food & Drug Research Laboratories
Hazleton Laboratories America, Inc.
IIT Research Institute
Laboratory Research Enterprises, Inc.
M. B. Research Laboratories, Inc.
Microbiological Associates, Inc.
Midwest Research Institute
Research Triangle Institute
Ricerca, Inc.
Southern Research Institute
Southwest Research Institute
Springborn Life Sciences, Inc., Toxicology & Human Safety Division
SRI International, Life Sciences Division
Toxikon Corporation

TPS, Inc.
White Eagle Toxicology Laboratories

CYTOTOXICITY

Bushy Run Research Center

DERMATOLOGY EFFICACY

Education & Research Foundation, Inc.
Essex Testing
Hazleton Laboratories America, Inc.
Product Safety Laboratories

DEVELOPMENTAL TOXICOLOGY AND TERATOLOGY

Argus Research Laboratories, Inc.
Battelle Memorial Institute
Bushy Run Research Center
Dawson Research Corporation
Hazleton Laboratories America, Inc.
Microbiological Associates, Inc.
Research Triangle Institute
South Mountain Laboratories, Inc.
Springborn Life Sciences, Inc., Toxicology & Human Safety Division
Stillmeadow, Inc.
White Eagle Toxicology Laboratories

ECOTOXICOLOGY/ENVIRONMENTAL FATE

Analytical Bio-Chemistry Laboratories, Inc.
Arthur D. Little, Chemical & Life Sciences Section
Battelle Memorial Institute
Bio-Life Associates Ltd.
BIOMED, Inc.
ECS/Normandeau
Enseco, Inc.
Harris Laboratories, Inc.
Hazleton Laboratories America, Inc.
Hunter Environmental Services, Inc.
Langston Laboratories, Inc.
Lycott Environmental Research, Inc.
Malcolm Pirnie, Inc.
Microbiological Associates, Inc.
Midwest Research Institute
Oak Creek Laboratory of Biology, Department of Fisheries & Wildlife
Ricerca, Inc.
Roy F. Weston, Inc.
Savannah Laboratories & Environmental Services, Inc.
Southwest Research Institute

APPENDIX B: LIST OF STUDIES CONDUCTED AND/OR SERVICES PROVIDED 93

Springborn Life Sciences, Inc., Environmental Toxicology & Chemistry Division
Syracuse Research Corporation
Thermo Analytical, Inc.
Wildlife International Ltd.

GENOTOXICITY

American Health Foundation
Applied Genetics
Arthur D. Little, Chemical & Life Sciences Section
BIOMED, Inc.
Bionetics Research, Inc.
Bushy Run Research Center
Hazleton Laboratories America, Inc.
IIT Research Institute
Microbiological Associates, Inc.
Midwest Research Institute
North American Science Associates, Inc.
Product Safety Laboratories
Research Triangle Institute
Southern Research Institute
SRI International, Life Sciences Division
Stillmeadow, Inc.
Toxikon Corporation

ENVIRONMENTAL FATE

Analytical Bio-Chemistry Laboratories, Inc.
Battelle Memorial Institute
Hazleton Laboratories America, Inc.
Lancaster Laboratories, Inc.
Toxikon Corporation

IMMUNOTOXICOLOGY

Arthur D. Little, Chemical & Life Sciences Section
Colorado Histo-Prep, Inc.
IIT Research Institute
ImmuQuest Laboratories, Inc.
Northview Pacific Laboratories, Inc.

INDUSTRIAL HYGIENE/ANALYTICAL

Kemron Environmental Services
Lancaster Laboratories, Inc.
Nebraska Testing Corporation

INHALATION TOXICOLOGY

Arthur D. Little, Chemical & Life Sciences Section
BIOMED, Inc.

Bushy Run Research Center
Food & Drug Research Laboratories
Hazleton Laboratories America, Inc.
IIT Research Institute
Leberco Testing, Inc.
Microbiological Associates, Inc.
Product Safety Laboratories
Springborn Life Sciences, Inc., Toxicology & Human Safety Division
Stillmeadow, Inc.

METABOLISM

Arthur D. Little, Chemical & Life Sciences Section
Analytical Bio-Chemistry Laboratories, Inc.
Battelle Memorial Institute
Bio/dynamics, Inc.
Biological Test Center
BIOMED, Inc.
Comparative Toxicology Laboratories, Kansas State University
Harris Laboratories, Inc.
Hazleton Laboratories America, Inc.
Midwest Research Institute
Ricerca, Inc.
Springborn Life Sciences, Inc., Environmental Toxicology & Chemistry Division
Wildlife International Ltd.

MICROBIOLOGY

Langston Laboratories, Inc.
La Rocca Science Laboratories
Microbiological Associates, Inc.
Nebraska Testing Corporation
Scientific Associates, Inc.

NEUROTOXICOLOGY (BEHAVIORAL AND NEUROBEHAVIORAL)

Battelle Memorial Institute
Bushy Run Research Center
Hazleton Laboratories America, Inc.
Research Triangle Institute
Southwest Research Institute
SRI International, Life Sciences Division

PHARMACOKINETICS/TOXICOKINETICS

Arthur D. Little, Chemical & Life Sciences Section
Battelle Memorial Institute
Biological Test Center
Bushy Run Research Center

Hazleton Laboratories America, Inc.
IIT Research Institute
Midwest Research Institute
Research Triangle Institute
Southwest Research Institute
SRI International, Life Sciences
    Division

RADIOLABEL

American Radiolabeled Chemicals, Inc.
Amersham Corporation
Chemsyn Science Laboratories,
    Member of the Specialty Materials
    Division,
    Eagle-Picher Industries, Inc.
Wizard Laboratories, Inc.

REPRODUCTION

Argus Research Laboratories, Inc.
Arthur D. Little, Chemical & Life Sciences
    Section
Battelle Memorial Institute
Bio-Life Associates Ltd.
Hazleton Laboratories America, Inc.
Laboratory Research Enterprises, Inc.
Microbiological Associates, Inc.
Oak Creek Laboratory of Biology, Department
    of Fisheries & Wildlife
Research Triangle Institute
Ricerca, Inc.
Springborn Life Sciences, Inc., Toxicology &
    Human Safety Division
SRI International, Life Sciences Division
White Eagle Toxicology Laboratories
Wildlife International Ltd.

SPECIAL STUDIES IN TOXICOLOGY

Ani Lytics, Inc.
Bushy Run Research Center
Colorado Histo-Prep, Inc.
Education & Research Foundation, Inc.
Hazards Research Corporation
Hazleton Laboratories America, Inc.
Stillmeadow, Inc.

SUBCHRONIC

Arthur D. Little, Chemical & Life Sciences
    Section
Battelle Memorial Institute
Bio/dynamics, Inc.
Biological Test Center
BIOMED, Inc.
Bushy Run Research Center
Comparative Toxicology Laboratories, Kansas
    State University
Dawson Research Corporation
Enseco, Inc.
Food & Drug Research Laboratories
Hazleton Laboratories America, Inc.
IIT Research Institute
Laboratory Research Enterprises, Inc.
Leberco Testing, Inc.
M. B. Research Laboratories, Inc.
Microbiological Associates, Inc.
Midwest Research Institute
Product Safety Laboratories
Ricerca, Inc.
Scientific Associates, Inc.
Southwest Research Institute
Springborn Life Sciences, Inc., Toxicology &
    Human Safety Division
SRI International, Life Sciences Division
Toxikon Corporation
TPS, Inc.
White Eagle Toxicology Laboratories

Appendix C

Types of Chemicals Tested

AGROCHEMICAL TESTING

Ana-Lab Corporation, 2600 Dudley Rd., Kilgore, TX 75662
Analytical Bio-Chemistry Laboratories, Inc., 7200 East ABC Lane, Columbia, MO 65202
Arthur D. Little, Chemical & Life Sciences Section, 25 Acorn Park, Cambridge, MA 02140
Arthur D. Little, Chemical & Life Sciences Section, 30 Memorial Dr., Cambridge, MA 02140
Bio/dynamics, Inc., Mettlers Rd., Box 2360, East Millstone, NJ 08875-2360
Hazleton Laboratories America, Inc., 3301 Kinsman Blvd., Madison, WI 53704
Microbiological Associates, Inc., 9900 Blackwell Rd., Rockville, MD 20850
SRI International, Life Sciences Division, 333 Ravenswood Ave., Menlo Park, CA 94025-3493

AIR TESTING

Ana-Lab Corporation, 2600 Dudley Rd., Kilgore, TX 75662
Battelle Memorial Institute, 505 King Ave., Columbus, OH 43201-2693
Biospherics, Inc., 12051 Indian Creek Court, Beltsville, MD 12051
Hazleton Laboratories America, Inc., 900 Osceola Dr., West Palm Beach, FL 33409
La Rocca Science Laboratories, Inc., 1 Nell Court, Dumont, NJ 07628
Lancaster Laboratories, Inc., 2425 New Holland Pike, Lancaster, PA 17601
Roy F. Weston, Inc., Weston Way, West Chester, PA 19380
Serco Laboratories, 1931 Westcounty Rd. C-2, St. Paul, MN 55113
Southern Research Institute, 2000 Ninth Ave., South, Box 55305, Birmingham, AL 35255-5305
Structure Probe, Inc., 1015 Merrick Rd., Copiague, NY 11726
Thermo Analytical, Inc., 160 Taylor St., Monrovia, CA 91016

ASBESTOS TESTING

Biospherics, Inc., 12051 Indian Creek Court, Beltsville, MD 12051
Lancaster Laboratories, Inc., 2425 New Holland Pike, Lancaster, PA 17601

BIOCIDES TESTING

Ricerca, Inc., 7528 Auburn Rd., Painesville, OH 44077

CERAMICS TESTING

Structure Probe, Inc., Box 656, West Chester, PA 19381-9656
Structure Probe, Inc., 230 Forrest St., Metuchen, NJ 08840
Structure Probe, Inc., 63 Unquowa Rd., Fairfield, CT 06430

CHEMICALS TESTING

American Health Foundation, 1 Dana Rd., Valhalla, NY 10595
Analytical Bio-Chemistry Laboratories, Inc., 7200 East ABC Lane, Columbia, MO 65202
Argus Research Laboratories, Inc., 935 Horsham Rd., Horsham, PA 19044
Arthur D. Little, Chemical & Life Sciences Section, 25 Acorn Park, Cambridge, MA 02140
Arthur D. Little, Chemical & Life Sciences Section, 30 Memorial Dr., Cambridge, MA 02140
Bio-Life Associates Ltd., Route 3, Box 156, Neillsville, WI 54456
Bio/dynamics, Inc., Mettlers Rd., Box 2360, East Millstone, NJ 08875-2360
Biological Test Center, 2525 McGaw Ave., Irvine, CA 92714
Bushy Run Research Center, R.D. 4, Mellon Rd., Export, PA 15632
Comparative Toxicology Laboratories VCS, Kansas State University, Manhattan, KS 66506
Cosmopolitan Safety Evaluation, Inc., Statesville Quarry Rd., Box 71, Lafayette, NJ 07848
Dawson Research Corporation, Box 620666, Orlando, FL 32862-0666
Enseco, Inc., Doaks Lane at Little Harbor, Marblehead, MA 01945
Essex Testing, 799 Bloomfield Ave., Verona, NJ 07044
Food and Drug Research Laboratories, Division of Enviro/Analysis Corporation, Rt. 17C, Box 107, Waverly, NY 14892
Hazards Research Corporation, 200 East Main St., Rockaway, NJ 07866
Hazleton Laboratories America, Inc., 3301 Kinsman Blvd., Madison, WI 53704
Hunter Environmental Services, Inc., Box 1703, Gainesville, FL 32602-1703
IIT Research Institute, 10 W. 35th St., Chicago, IL 60616-3799
Laboratory Research Enterprises, Inc., 6321 South 6th St., Kalamazoo, MI 49009-9611
Lancaster Laboratories, Inc., 2425 New Holland Pike, Lancaster, PA 17601
Leberco Testing, Inc., 123 Hawthorne St., Roselle Park, NJ 07204-0206
Life Science Research, Ltd., Eye, Suffolk IP23 7PX, England
Microbiological Associates, Inc., 9900 Blackwell Rd., Rockville, MD 20850
Midwest Research Institute, 425 Volker Blvd., Kansas City, MO 64110
North American Science Associates, Inc., 2261 Tracy Rd., Northwood, OH 43619
North American Science Associates, Inc., 9 Morgan, Irvine, CA 92718
Northview Pacific Laboratories, Inc., 2800 Seventh St., Berkeley, CA 94710
Northview Pacific Laboratories, Inc., 1880 Holste Rd., Northbrook, IL 60062
Oak Creek Laboratory of Biology, Department of Fisheries & Wildlife, Nash Hall, Room #104, Oregon State University, Corvallis, OR 97331-3803
Product Safety Laboratories, 725 Cranbury Rd., East Brunswick, NJ 08816
Research Triangle Institute, 3040 Cornwallis Rd., Box 12194, Research Triangle Park, NC 27709
Ricerca, Inc., 7528 Auburn Rd., Painesville, OH 44077-1000
Scientific Associates, Inc., 6200 South Lindbergh Blvd., St. Louis, MO 63123
Springborn Life Sciences, Inc., Toxicology & Human Safety Division, 553 North Broadway, Spencerville, OH 45887

APPENDIX C: TYPES OF CHEMICALS TESTED

Springborn Life Sciences, Inc., Toxicology & Human Safety Division, 790 Main St., Wareham, MA 02543
SRI International, Life Sciences Division, 333 Ravenswood Ave., Menlo Park, CA 94025-3493
Stillmeadow, Inc., 9525 Town Park Dr., Houston, TX 77036
Syracuse Research Corporation, Merrill Lane, Syracuse, NY 13210-4080
Toxikon Corporation, 225 Wildwood Ave., Woburn, MA 01801
White Eagle Toxicology Laboratories, 2003 Lower State Rd., Doylestown, PA 18901
Wildlife International Ltd., 305 Commerce Dr., Easton, MD 21601

CONSUMER PRODUCTS TESTING

Arthur D. Little, Chemical & Life Sciences Section, 25 Acorn Park, Cambridge, MA 02140
Arthur D. Little, Chemical & Life Sciences Section, 30 Memorial Dr., Cambridge, MA 02140
Bio/dynamics, Inc., Mettlers Rd., Box 2360, East Millstone, NJ 08875-2360
Harris Laboratories, Inc., 624 Peach St., Box 80837, Lincoln, NB 68501
Microbiological Associates, Inc., 9900 Blackwell Rd., Rockville, MD 20850
SRI International, Life Sciences Division, 333 Ravenswood Ave., Menlo Park, CA 94025-3493

COSMETIC/TOILETRIES TESTING

Analytical Bio-Chemistry Laboratories, Inc., 7200 East ABC Lane, Columbia, MO 65202
Argus Research Laboratories, Inc., 935 Horsham Rd., Horsham, PA 19044
Arthur D. Little, Chemical & Life Sciences Section, 25 Acorn Park, Cambridge, MA 02140
Arthur D. Little, Chemical & Life Sciences Section, 30 Memorial Dr., Cambridge, MA 02140
Bio-Life Associates Ltd., Route 3, Box 156, Neillsville, WI 54456
Bio/dynamics, Inc., Mettlers Rd., Box 2360, East Millstone, NJ 08875-2360
Biological Test Center, 2525 McGaw Ave., Irvine, CA 92714
BIOMED, Inc., 1720 130th Ave., N.E., Bellevue, WA 98005-2203
Bushy Run Research Center, R.D. 4, Mellon Rd., Export, PA 15632
Cosmopolitan Safety Evaluation, Inc., Statesville Quarry Rd., Box 71, Lafayette, NJ 07848
Dawson Research Corporation, Box 620666, Orlando, FL 32862-0666
Education & Research Foundation, Inc., 2602 Langhorne Rd., Lynchburg, VA 24501
Essex Testing, 799 Bloomfield Ave., Verona, NJ 07044
Food and Drug Research Laboratories Division of Enviro/Analysis Corporation, Rt. 17C, Box 107, Waverly, NY 14892
Hazleton Laboratories America, Inc., 900 Osceola Dr., West Palm Beach, FL 33409
ImmuQuest Laboratories, Inc., 13 Taft Court, Suite 200, Rockville, MD 20850
La Rocca Science Laboratories, Inc., 1 Nell Court, Dumont, NJ 07628
Lancaster Laboratories, Inc., 2425 New Holland Pike, Lancaster, PA 17601
Leberco Testing, Inc., 123 Hawthorne St., Roselle Park, NJ 07204-0206
North American Science Associates, Inc., 2261 Tracy Rd., Northwood, OH 43619
North American Science Associates, Inc., 9 Morgan, Irvine, CA 92718
Northview Pacific Laboratories, Inc., 2800 Seventh St., Berkeley, CA 94710
Northview Pacific Laboratories, Inc., 1880 Holste Rd., Northbrook, IL 60062
Product Safety Laboratories, 725 Cranbury Rd., East Brunswick, NJ 08816
Research Triangle Institute, 3040 Cornwallis Rd., Box 12194, Research Triangle Park, NC 27709
Ricerca, Inc., 7528 Auburn Rd., Painesville, OH 44077
Scientific Associates, Inc., 6200 South Lindbergh Blvd., St. Louis, MO 63123
South Mountain Laboratories, Inc., 380 Lackawanna Place, South Orange, NJ 07079
Stillmeadow, Inc., 9525 Town Park Dr., Houston, TX 77036
Structure Probe, Inc., Box 656, West Chester, PA 19381-9656
Structure Probe, Inc., 230 Forrest St., Metuchen, NJ 08840
Structure Probe, Inc., 63 Unquowa Rd., Fairfield, CT 06430
Toxikon Corporation, 225 Wildwood Ave., Woburn, MA 01801

DRUG TESTING

Bionetics Research, Inc., 5516 Nicholson Lane, Kensington, MD 20895-1078
Comparative Toxicology Laboratories VCS, Kansas State University, Manhattan, KS 66506
Hazards Research Corporation, 200 East Main St., Rockaway, NJ 07866
Midwest Research Institute, 425 Volker Blvd., Kansas City, MO 64110
Northview Pacific Laboratories, Inc., 2800 Seventh St., Berkeley, CA 94710
Northview Pacific Laboratories, Inc., 1880 Holste Rd., Northbrook, IL 60062
Research Triangle Institute, 3040 Cornwallis Rd., Box 12194, Research Triangle Park, NC 27709
Springborn Life Sciences, Inc., Toxicology & Human Safety Division, 553 North Broadway, Spencerville, OH 45887
Springborn Life Sciences, Inc., Toxicology & Human Safety Division, 790 Main St., Wareham, MA 02543

EFFLUENTS TESTING

Analytical Bio-Chemistry Laboratories, Inc., 7200 East ABC Lane, Columbia, MO 65202
Enseco, Inc., Doaks Lane at Little Harbor, Marblehead, MA 01945
Hunter Environmental Services, Inc., Box 1703, Gainesville, FL 32602-1703
Malcolm Pirnie, Inc., 2 Corporate Park Dr., White Plains, NY 10602
Malcolm Pirnie, Inc., 100 Grasslands Rd., Elmsford, NY 10523
Southern Research Institute, 2000 Ninth Ave., South, Box 55305, Birmingham, AL 35255-5305
Springborn Life Sciences, Inc., Toxicology & Human Safety Division, 790 Main St., Wareham, MA 02543
Wildlife International Ltd., 305 Commerce Dr., Easton, MD 21601

ELASTOMERS TESTING

Structure Probe, Inc., Box 656, West Chester, PA 19381-9656
Structure Probe, Inc., 230 Forrest St., Metuchen, NJ 08840
Structure Probe, Inc., 63 Unquowa Rd., Fairfield, CT 06430

HAZARDOUS CHEMICALS/WASTE TESTING

BIOMED, Inc., 1720 130th Ave., N.E., Bellevue, WA 98005-2203
ECS/Normandeau, Box 1393, Aiken, SC 29802
Serco Laboratories, 1931 Westcounty Rd. C-2, St. Paul, MN 55113
Southern Research Institute, 2000 Ninth Ave., South, Box 55305, Birmingham, AL 35255-5305
Thermo Analytical, Inc., 160 Taylor St., Monrovia, CA 91016

HERBICIDES TESTING

Leberco Testing, Inc., 123 Hawthorne St., Roselle Park, NJ 07204-0206
Ricerca, Inc., 7528 Auburn Rd., Painesville, OH 44077

HOUSEHOLD PRODUCTS TESTING

Hazleton Laboratories America, Inc., 3301 Kinsman Blvd., Madison, WI 53704
Leberco Testing, Inc., 123 Hawthorne St., Roselle Park, NJ 07204-0206
Product Safety Laboratories, 725 Cranbury Rd., East Brunswick, NJ 08816

APPENDIX C: TYPES OF CHEMICALS TESTED

INDUSTRIAL CHEMICALS TESTING

Analytical Bio-Chemistry Laboratories, Inc., 7200 East ABC Lane, Columbia, MO 65202
Bio-Life Associates Ltd., Route 3, Box 156, Neillsville, WI 54456
Bushy Run Research Center, R.D. 4, Mellon Rd., Export, PA 15632
Cosmopolitan Safety Evaluation, Inc., Statesville Quarry Rd., Box 71, Lafayette, NJ 07848
ECS/Normandeau, Box 1393, Aiken, SC 29802
Enseco, Inc., Doaks Lane at Little Harbor, Marblehead, MA 01945
Food and Drug Research Laboratories, Division of Enviro/Analysis Corporation, Rt. 17C, Box 107, Waverly, NY 14892
Hazards Research Corp., 200 East Main St., Rockaway, NJ 07866
Hazleton Laboratories America, Inc., 3301 Kinsman Blvd., Madison, WI 53704
Hunter Environmental Services, Inc., Box 1703, Gainesville, FL 32602-1703
IIT Research Institute, 10 W. 35th St., Chicago, IL 60616-3799
Laboratory Research Enterprises, Inc., 6321 South 6th St., Kalamazoo, MI 49009-9611
Lancaster Laboratories, Inc., 2425 New Holland Pike, Lancaster, PA 17601
Leberco Testing, Inc., 123 Hawthorne St., Roselle Park, NJ 07204-0206
M. B. Research Laboratories, Inc., Steinberg & Wentz Rds., Box 178, Spinnerstown, PA 18968
Microbiological Associates, Inc., 9900 Blackwell Rd., Rockville, MD 20850
Midwest Research Institute, 425 Volker Blvd., Kansas City, MO 64110
Oak Creek Laboratory of Biology, Department of Fisheries & Wildlife, Nash Hall, Room #104, Oregon State University, Corvallis, OR 97331-3803
Product Safety Laboratories, 725 Cranbury Rd., East Brunswick, NJ 08816
Research Triangle Institute, 3040 Cornwallis Rd., Box 12194, Research Triangle Park, NC 27709
Ricerca, Inc., 7528 Auburn Rd., Painesville, OH 44077
Scientific Associates, Inc., 6200 South Lindbergh Blvd., St. Louis, MO 63123
Southwest Research Institute, 6220 Culebra Road, Drawer 28510, San Antonio, TX 78284-2900
Springborn Life Sciences, Inc., Toxicology & Human Safety Division, 553 North Broadway, Spencerville, OH 45887
Springborn Life Sciences, Inc., Toxicology & Human Safety Division, 790 Main St., Wareham, MA 02543
SRI International, Life Sciences Division, 333 Ravenswood Ave., Menlo Park, CA 94025-3493
Stillmeadow, Inc., 9525 Town Park Dr., Houston, TX 77036
Syracuse Research Corp., Merrill Lane, Syracuse, NY 13210-4080
Toxikon Corporation, 225 Wildwood Ave., Woburn, MA 01801
White Eagle Toxicology Laboratories, 2003 Lower State Rd., Doylestown, PA 18901
Wildlife International Ltd., 305 Commerce Dr., Easton, MD 21601

LUBRICANTS TESTING

IIT Research Institute, 10 W. 35th St., Chicago, IL 60616-3799

MEDICAL DEVICES TESTING

Argus Research Laboratories, Inc., 935 Horsham Rd., Horsham, PA 19044
Biological Test Center, 2525 McGaw Ave., Irvine, CA 92714
Hazleton Laboratories America, Inc., 3301 Kinsman Blvd., Madison, WI 53704
Leberco Testing, Inc., 123 Hawthorne St., Roselle Park, NJ 07204-0206
Microbiological Associates, Inc., 9900 Blackwell Rd., Rockville, MD 20850
North American Science Associates, Inc., 2261 Tracy Rd., Northwood, OH 43619
North American Science Associates, Inc., 9 Morgan, Irvine, CA 92718
Northview Pacific Laboratories, Inc., 2800 Seventh St., Berkeley, CA 94710
Northview Pacific Laboratories, Inc., 1880 Holste Rd., Northbrook, IL 60062
SRI International, Life Sciences Division, 333 Ravenswood Ave., Menlo Park, CA 94025-3493
Stillmeadow, Inc., 9525 Town Park Dr., Houston, TX 77036
Toxikon Corporation, 225 Wildwood Ave., Woburn, MA 01801

METALS TESTING

Comparative Toxicology Laboratories VCS, Kansas State University, Manhattan, KS 66506
Ecology & Environment, Inc., Analytical Services Center, 4285 Genesee St., Buffalo, NY 14225
ECS/Normandeau, Box 1393, Aiken, SC 29802
Hazards Research Corp., 200 East Main St., Rockaway, NJ 07866
Hazleton Laboratories America, Inc., 900 Osceola Dr., West Palm Beach, FL 33409
Lancaster Laboratories, Inc., 2425 New Holland Pike, Lancaster, PA 17601
Midwest Research Institute, 425 Volker Blvd., Kansas City, MO 64110
Northview Pacific Laboratories, Inc., 2800 Seventh St., Berkeley, CA 94710
Northview Pacific Laboratories, Inc., 1880 Holste Rd., Northbrook, IL 60062
Recon Systems, Inc., Route 202 North, Box 460, Three Bridges, NJ 08887
Research Triangle Institute, 3040 Cornwallis Rd., Box 12194, Research Triangle Park, NC 27709
Savannah Laboratories & Environmental Services, Inc.
    Savannah Div., 5102 LaRoche Ave., Savannah, GA 31404
    Mobile Div., 3707 Cottage Hill Rd., Mobile, AL 36609
    Tallahassee Div., 2820 Industrial Plaza, Tallahassee, FL 32301
Structure Probe, Inc., Box 656, West Chester, PA 19381-9656
Structure Probe, Inc., 230 Forrest St., Metuchen, NJ 08840
Structure Probe, Inc., 63 Unquowa Rd., Fairfield, CT 06430
Syracuse Research Corp., Merrill Lane, Syracuse, NY 13210-4080

MICROBIOLOGICAL TESTING

Biospherics, Inc., 12051 Indian Creek Court, Beltsville, MD 12051
Cosmopolitan Safety Evaluation, Inc., Statesville Quarry Rd., Box 71, Lafayette, NJ 07848
Lancaster Laboratories, Inc., 2425 New Holland Pike, Lancaster, PA 17601
Scientific Associates, Inc., 6200 South Lindbergh Blvd., St. Louis, MO 63123
Springborn Life Sciences, Inc., Toxicology & Human Safety Division, 790 Main St., Wareham, MA 02543

PESTICIDE TESTING

Argus Research Laboratories, Inc., 935 Horsham Rd., Horsham, PA 19044
Battelle Memorial Institute, 505 King Ave., Columbus, OH 43201-2693
Bio-Life Associates Ltd., Route 3, Box 156, Neillsville, WI 54456
Biological Test Center, 2525 McGaw Ave., Irvine, CA 92714
Bushy Run Research Center, R.D. 4, Mellon Rd., Export, PA 15632
Cosmopolitan Safety Evaluation, Inc., Statesville Quarry Rd., Box 71, Lafayette, NJ 07848
Dawson Research Corp., Box 620666, Orlando, FL 32862-0666
Enseco, Inc., Doaks Lane at Little Harbor, Marblehead, MA 01945
Food and Drug Research Laboratories, Division of Enviro/Analysis Corporation, Rt. 17C, Box 107, Waverly, NY 14892
Hazleton Laboratories America, Inc., 3301 Kinsman Blvd., Madison, WI 53704
Hunter Environmental Services, Inc., Box 1703, Gainesville, FL 32602-1703
IIT Research Institute, 10 W. 35th St., Chicago, IL 60616-3799
ImmuQuest Laboratories, Inc., 13 Taft Court, Suite 200, Rockville, MD 20850
Lancaster Laboratories, Inc., 2425 New Holland Pike, Lancaster, PA 17601
Leberco Testing, Inc., 123 Hawthorne St., Roselle Park, NJ 07204-0206
Midwest Research Institute, 425 Volker Blvd., Kansas City, MO 64110
Oak Creek Laboratory of Biology, Department of Fisheries & Wildlife, Nash Hall, Room #104, Oregon State University, Corvallis, OR 97331-3803
Product Safety Laboratories, 725 Cranbury Rd., East Brunswick, NJ 08816
Ricerca, Inc., 7528 Auburn Rd., Painesville, OH 44077
Scientific Associates, Inc., 6200 South Lindbergh Blvd., St. Louis, MO 63123
Southwest Research Institute, 6220 Culebra Road, Drawer 28510, San Antonio, TX 78284-2900

APPENDIX C: TYPES OF CHEMICALS TESTED

Springborn Life Sciences, Inc., Toxicology & Human Safety Division, 553 North Broadway, Spencerville, OH 45887
Springborn Life Sciences, Inc., Toxicology & Human Safety Division, 790 Main St., Wareham, MA 02543
SRI International, Life Sciences Division, 333 Ravenswood Ave., Menlo Park, CA 94025-3493
Stillmeadow, Inc., 9525 Town Park Dr., Houston, TX 77036
Toxikon Corporation, 225 Wildwood Ave., Woburn, MA 01801
White Eagle Toxicology Laboratories, 2003 Lower State Rd., Doylestown, PA 18901
Wildlife International Ltd., 305 Commerce Dr., Easton, MD 21601

PET FOODS TESTING

Arthur D. Little, Chemical & Life Sciences Section, 25 Acorn Park, Cambridge, MA 02140
Arthur D. Little, Chemical & Life Sciences Section, 30 Memorial Dr., Cambridge, MA 02140
IIT Research Institute, 10 W. 35th St., Chicago, IL 60616-3799
Laboratory Research Enterprises, Inc., 6321 South 6th St., Kalamazoo, MI 49009-9611
Microbiological Associates, Inc., 9900 Blackwell Rd., Rockville, MD 20850
SRI International, Life Sciences Division, 333 Ravenswood Ave., Menlo Park, CA 94025-3493

PETROLEUM TESTING

Microbiological Associates, Inc., 9900 Blackwell Rd., Rockville, MD 20850
SRI International, Life Sciences Division, 333 Ravenswood Ave., Menlo Park, CA 94025-3493

PHARMACEUTICALS TESTING

Analytical Bio-Chemistry Laboratories, Inc., 7200 East ABC Lane, Columbia, MO 65202
Ani Lytics, Inc., 360 Christopher Ave., Gaithersburg, MD 20879
Applied Genetics, Inc., 1335 Gateway Dr. #2001, Melbourne, FL 32901-2619
Argus Research Laboratories, Inc., 935 Horsham Rd., Horsham, PA 19044
Arthur D. Little, Chemical & Life Sciences Section, 25 Acorn Park, Cambridge, MA 02140
Arthur D. Little, Chemical & Life Sciences Section, 30 Memorial Dr., Cambridge, MA 02140
Bio-Life Associates Ltd., Route 3, Box 156, Neillsville, WI 54456
Bio/dynamics, Inc., Mettlers Rd., Box 2360, East Millstone, NJ 08875-2360
Biological Test Center, 2525 McGaw Ave., Irvine, CA 92714
BIOMED, Inc., 1720 130th Ave., N.E., Bellevue, WA 98005-2203
Bushy Run Research Center, R.D. 4, Mellon Rd., Export, PA 15632
Cosmopolitan Safety Evaluation, Inc., Statesville Quarry Rd., Box 71, Lafayette, NJ 07848
Dawson Research Corp., Box 620666, Orlando, FL 32862-0666
Education & Research Foundation, Inc., 2602 Langhorne Rd., Lynchburg, VA 24501
Essex Testing, 799 Bloomfield Ave., Verona, NJ 07044
Food and Drug Research Laboratories, Division of Enviro/Analysis Corporation, Rt. 17C, Box 107, Waverly, NY 14892
Harris Laboratories, Inc., 624 Peach St., Box 80837, Lincoln, NB 68501
Hazleton Laboratories America, Inc., 3301 Kinsman Blvd., Madison, WI 53704
900 Osceola Dr., West Palm Beach, FL 33409
IIT Research Institute, 10 W. 35th St., Chicago, IL 60616-3799
La Rocca Science Laboratories, Inc., 1 Nell Court, Dumont, NJ 07628
Laboratory Research Enterprises, Inc., 6321 South 6th St., Kalamazoo, MI 49009-9611
Lancaster Laboratories, Inc., 2425 New Holland Pike, Lancaster, PA 17601
Leberco Testing, Inc., 123 Hawthorne St., Roselle Park, NJ 07204-0206
Microbiological Associates, Inc., 9900 Blackwell Rd., Rockville, MD 20850
North American Science Associates, Inc., 2261 Tracy Rd., Northwood, OH 43619
North American Science Associates, Inc., 9 Morgan, Irvine, CA 92718
Northview Pacific Laboratories, Inc., 2800 Seventh St., Berkeley, CA 94710
Northview Pacific Laboratories, Inc., 1880 Holste Rd., Northbrook, IL 60062

Product Safety Laboratories, 725 Cranbury Rd., East Brunswick, NJ 08816
Research Triangle Institute, 3040 Cornwallis Rd., Box 12194, Research Triangle Park, NC 27709
Ricerca, Inc., 7528 Auburn Rd., Painesville, OH 44077
Scientific Associates, Inc., 6200 South Lindbergh Blvd., St. Louis, MO 63123
South Mountain Laboratories, Inc., 380 Lackawanna Place, South Orange, NJ 07079
Southwest Research Institute, 6220 Culebra Road, Drawer 28510, San Antonio, TX 78284-2900
Springborn Life Sciences, Inc., Toxicology & Human Safety Division, 553 North Broadway, Spencerville, OH 45887
SRI International, Life Sciences Division, 333 Ravenswood Ave., Menlo Park, CA 94025-3493
Stillmeadow, Inc., 9525 Town Park Dr., Houston, TX 77036
Structure Probe, Inc., Box 656, West Chester, PA 19381-9656
Structure Probe, Inc., 230 Forrest St., Metuchen, NJ 08840
Structure Probe, Inc., 63 Unquowa Rd., Fairfield, CT 06430
Syracuse Research Corp., Merrill Lane, Syracuse, NY 13210-4080
Toxikon Corporation, 225 Wildwood Ave., Woburn, MA 01801
White Eagle Toxicology Laboratories, 2003 Lower State Rd., Doylestown, PA 18901
Wildlife International Ltd., 305 Commerce Dr., Easton, MD 21601

PLASTICS TESTING

Ana-Lab Corporation, 2600 Dudley Rd., Kilgore, TX 75662
BIOMED, Inc., 1720 130th Ave., N.E., Bellevue, WA 98005-2203
Biospherics, Inc., 12051 Indian Creek Court, Beltsville, MD 12051
ECS/Normandeau, Box 1393, Aiken, SC 29802
Enseco, Inc., Doaks Lane at Little Harbor, Marblehead, MA 01945
Enviroscan, Inc., 303 W. Military Rd., Rothschild, WI 54474
Hazards Research Corp., 200 East Main St., Rockaway, NJ 07866
Leberco Testing, Inc., 123 Hawthorne St., Roselle Park, NJ 07204-0206
Malcolm Pirnie, Inc., 2 Corporate Park Dr., White Plains, NY 10602
Malcolm Pirnie, Inc., 100 Grasslands Rd., Elmsford, NY 10523
Roy F. Weston, Inc., Weston Way, West Chester, PA 19380
Serco Laboratories, 1931 Westcounty Rd. C-2, St. Paul, MN 55113
Southern Research Institute, 2000 Ninth Ave., South, Box 55305, Birmingham, AL 35255-5305
Southwest Research Institute, 6220 Culebra Road, Drawer 28510, San Antonio, TX 78284-2900
Springborn Life Sciences, Inc., Toxicology & Human Safety Division, 790 Main St., Wareham, MA 02543
Stillmeadow, Inc., 9525 Town Park Dr., Houston, TX 77036
Structure Probe, Inc., Box 656, West Chester, PA 19381-9656
Structure Probe, Inc., 230 Forrest St., Metuchen, NJ 08840
Structure Probe, Inc., 63 Unquowa Rd., Fairfield, CT 06430
Thermo Analytical, Inc., 160 Taylor St., Monrovia, CA 91016

SOIL TESTING

Battelle Memorial Institute, 505 King Ave., Columbus, OH 43201-2693
Biospherics, Inc., 12051 Indian Creek Court, Beltsville, MD 12051
Enviroscan, Inc., 303 W. Military Rd., Rothschild, WI 54474
Harris Laboratories, Inc., 624 Peach St., Box 80837, Lincoln, NB 68501
Roy F. Weston, Inc., Weston Way, West Chester, PA 19380
Serco Laboratories, 1931 Westcounty Rd. C-2, St. Paul, MN 55113
Skinner & Sherman Laboratories, Inc., 300 Second Ave., Box 521, Waltham, MA 02254
Springborn Life Sciences, Inc., Toxicology & Human Safety Division, 790 Main St., Wareham, MA 02543
Thermo Analytical, Inc., 160 Taylor St., Monrovia, CA 91016

SPECIALTY CHEMICALS TESTING

Essex Testing, 799 Bloomfield Ave., Verona, NJ 07044
North American Science Associates, Inc., 2261 Tracy Rd., Northwood, OH 43619
North American Science Associates, Inc., 9 Morgan, Irvine, CA 92718
Northview Pacific Laboratories, Inc., 2800 Seventh St., Berkeley, CA 94710
Northview Pacific Laboratories, Inc., 1880 Holste Rd., Northbrook, IL 60062

WATER/WASTEWATER TESTING

Ana-Lab Corporation, 2600 Dudley Rd., Kilgore, TX 75662
Battelle Memorial Institute, 505 King Ave., Columbus, OH 43201-2693
BIOMED, Inc., 1720 130th Ave., N.E., Bellevue, WA 98005-2203
Biospherics, Inc., 12051 Indian Creek Court, Beltsville, MD 12051
ECS/Normandeau, Box 1393, Aiken, SC 29802
Enviroscan, Inc., 303 W. Military Rd., Rothschild, WI 54474
Harris Laboratories, Inc., 624 Peach St., Box 80837, Lincoln, NB 68501
La Rocca Science Laboratories, Inc., 1 Nell Court, Dumont, NJ 07628
Lancaster Laboratories, Inc., 2425 New Holland Pike, Lancaster, PA 17601
Leberco Testing, Inc., 123 Hawthorne St., Roselle Park, NJ 07204-0206
Malcolm Pirnie, Inc., 2 Corporate Park Dr., White Plains, NY 10602
Malcolm Pirnie, Inc., 100 Grasslands Rd., Elmsford, NY 10523
Roy F. Weston, Inc., Weston Way, West Chester, PA 19380
Serco Laboratories, 1931 Westcounty Rd. C-2, St. Paul, MN 55113
Skinner & Sherman Laboratories, Inc., 300 Second Ave., Box 521, Waltham, MA 02254
Southern Research Institute, 2000 Ninth Ave., South, Box 55305, Birmingham, AL 35255-5305
Thermo Analytical, Inc., 160 Taylor St., Monrovia, CA 91016

831649